Angel Amado Recio Recio
Anais Marquez
Lenin Montaño

Biogas. Construction of biodigesters for farmers

Angel Amado Recio Recio
Anais Marquez
Lenin Montaño

Biogas. Construction of biodigesters for farmers

General notions for the design and construction of a biodigester

ScienciaScripts

Imprint

Any brand names and product names mentioned in this book are subject to trademark, brand or patent protection and are trademarks or registered trademarks of their respective holders. The use of brand names, product names, common names, trade names, product descriptions etc. even without a particular marking in this work is in no way to be construed to mean that such names may be regarded as unrestricted in respect of trademark and brand protection legislation and could thus be used by anyone.

Cover image: www.ingimage.com

This book is a translation from the original published under ISBN 978-620-2-13595-5.

Publisher:
Sciencia Scripts
is a trademark of
Dodo Books Indian Ocean Ltd. and OmniScriptum S.R.L publishing group

120 High Road, East Finchley, London, N2 9ED, United Kingdom
Str. Armeneasca 28/1, office 1, Chisinau MD-2012, Republic of Moldova, Europe
Printed at: see last page
ISBN: 978-620-6-49831-5

Content

Chapter 1

Introduction

Biogas is the gas produced by a metabolic process of decomposition of organic matter, animal manure, or the combination of both products and vegetable wastes by bacterial action without the presence of oxygen in the air. This process is called anaerobic digestion.

Anaerobic digestion is a process of degradation of organic matter in the absence of air (oxygen). This process is carried out by anaerobic microorganisms (thus being a biological process) acting inside a biodigester. This biodigester or reactor is nothing more than an airtight chamber, in which the organic matter is placed without oxygen so that the fermentation can take place.

Organic anaerobic digestion, in the absence of oxygen, and through the action of a group of spice bacteria, decomposes into gaseous products or "biogas" (CH4, CO2, H2, H2S, etc.), and into digestate, which is a mixture of mineral products (N, P, K, Ca, etc.) and diacyl degradation compounds (N, P, K, Ca, etc.).

Biogas contains a high percentage of methane, CH4 (between 50-70%), which makes it suitable for energy use by combustion in engines, turbines or boilers, either alone or mixed with other fuels. The controlled anaerobic digestion process is one of the most suitable processes for the reduction of greenhouse gas emissions, the energetic use of organic wastes and the maintenance and improvement of the fertilizer value of the treated products.

Anaerobic digestion can be applied, among others, to livestock and agricultural wastes, as well as to wastes from the processing industries of these products. These wastes can include slurry, manure, agricultural residues or crop surpluses, etc. These wastes can be treated independently or together in what is called co-digestion. Anaerobic digestion is also a suitable process for the treatment of wastewaters with a high organic load, such as those produced in many food industries.

1.2. - Brief history of biogas

When the Italian physicist Alessandro Volta first identified methane (CH4) as the flammable gas in bubbles emerging from swamps at the end of the 18th century, he could not have imagined how important this gas would become for human society in the centuries to come.

Interest in biogas as a viable energy resource has spread across the globe in the last two decades.

1600- The first mentions of biogas as it was identified by several scientists as a gas coming from the decomposition of organic matter.

1776-Volta discovers methane (CH4) in swamp gas.

1890 the first full-scale biodigester was built in the

India.

1896 in Exeter, England, street lamps were powered by gas collected from digesters fermenting the city's sewage sludge.

1869-For the first time biogas (methane) is used in a hospital in Bombay, India.

During the years of the Second World War, biodigesters began to spread in rural areas in Europe as well as in China and India, which became leaders in the field.

After the World Wars, the so-called biogas production plants began to spread in Europe, the product of which was used in tractors and automobiles of the time. The so-called Imhoff tanks for the treatment of collective sewage water spread all over the world.

The gas produced was used for the operation of the plants themselves, in municipal veluculas and in some cities it was even injected into the communal gas network. This diffusion was interrupted by the easy access to fossil fuels and it was not until the energy crisis of the 70's that the investigation and extension of gas production was restarted with great impetus all over the world, including most Latin American countries.

The last 20 years have been fruitful in terms of discoveries about the microbiological and biochemical process thanks to new laboratory equipment that allowed the study of the microorganisms involved in anaerobic conditions (absence of oxygen).

These advances in the understanding of the microbiological process have been accompanied by important achievements in applied research, resulting in major advances in the technological field.

Biogas is now used worldwide as a fuel source for both industrial and domestic use. Its exploitation has helped drive sustained economic development and has provided an alternative renewable energy source to coal and oil.

Agricultural activity and the proper management of rural waste can contribute significantly to the production and conversion of animal and vegetable waste (biomass) into different forms of energy. During the anaerobic digestion of biomass, through a series of biochemical reactions, biogas is generated, which is mainly constituted by methane (CH_4) and carbon dioxide (CO). This biogas can be captured and used as fuel and/or electricity. In this way, anaerobic digestion, as a method of waste treatment, allows to reduce the amount of polluting organic matter, stabilizing it (biofertilizers) and at the same time, to produce gaseous energy (biogas).

From a developed and developing country perspective, anaerobic biotechnology contributes to meeting three basic needs: a) improving sanitary conditions through pollution control; b) generation of renewable energy for domestic activities; and c) providing stabilized materials (biofertilizer) as a biofertilizer for crops. Therefore, anaerobic biotechnology plays an important role in pollution control and for obtaining valuable resources: energy and value-added products.

The most important technology-generating countries at present are: China, India, Holland, France, Great Britain, Switzerland, Italy, USA, Philippines and Germany.

Chapter 2 ANAEROBIC DIGESTION

2.1. -Biogas formation

The correct management of organic wastes is achieved through different treatments that involve the recycling of these organic materials, transforming them into value-added products. The recycling of organic matter has received a strong impulse with the high cost of chemical fertilizers, with the search for non-traditional energy alternatives, as well as the need for decontamination and waste disposal.

The microbial population plays an important role in the transformation of these organic wastes, especially if it is considered that they have a wide range of responses to the oxygen molecule, a universal component of the cells. This allows the establishment of bioprocesses depending on the presence or absence of oxygen, in order to adequately treat various organic wastes.

2.2. -Anaerobic Digestion

Anaerobic digestion is a complex and degradative biological process in which part of the organic materials of a substrate (animal and vegetable waste) are converted into biogas, a mixture of carbon dioxide and methane with traces of other elements, by a consortium of bacteria that are sensitive to or completely inhibited by oxygen or its precursors (e.g. H_2O_2). Using the anaerobic digestion process it is possible to convert large quantities of waste, vegetable waste, manure, effluents from the food and fermentation industry, the paper industry and some chemical industries, into useful by-products. In anaerobic digestion more than 90% of the energy available by direct oxidation is transformed into methane, consuming only 10% of the energy in bacterial growth compared to 50% consumed in an aerobic system.

It is a natural process that occurs spontaneously in nature and is part of the biological cycle. In this way we can find the so-called "swamp gas" that gushes from stagnant waters, the natural gas (methane) from oil fields, as well as the gas produced in the digestive tract of ruminants such as cattle. All these processes involve the so-called methanogenic bacteria.

In anaerobic digestion, methanogenic microorganisms play the role of respiratory enzymes and, together with non-methanogenic bacteria, constitute a food chain that is related to the enzyme chains of aerobic cells. In this way, the organic waste is completely transformed into biogas that leaves the system. However, the biogas generated is often contaminated with different components, which can complicate its handling and utilization.

The anaerobic process is classified as anaerobic fermentation or anaerobic respiration depending on the type of electron acceptors. Methanogenesis is a process that normally occurs in the rumen of herbivores and in other environments in the absence of oxygen, such as swamps, forest floor microenvironments or grasslands. Another important source of methane production is the decomposition of animal waste. With the exception of wood, which contains lignin, these anaerobic bacteria are capable of digesting practically any biological material. In this process carried out by bacteria, a mixture of gases is released. In Figure 2.1 and 2.2, you can see this process, the numbers indicate the bacterial population responsible for the process:

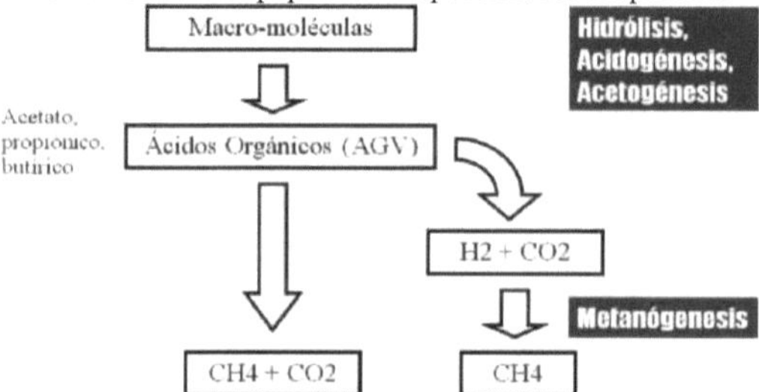

Fermentative bacteria
1. Hydrogen producing acetogenic bacteria
2. Homocetogenic bacteria
3. Methanogenic hydrogenotrophic bacteria

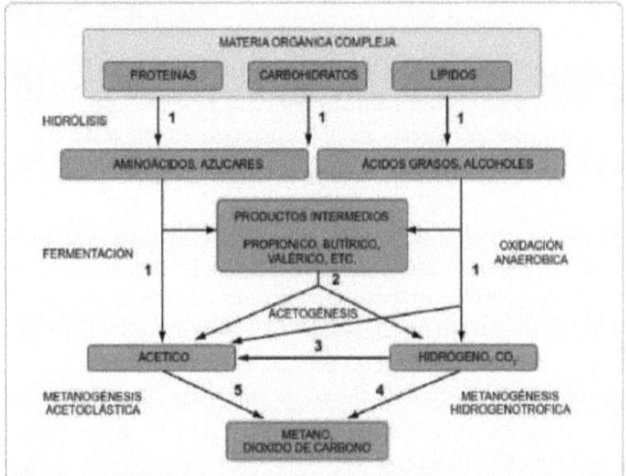

Fig.2.2.- Process of methanogenesis

The promotion and implementation of collective biogas production systems (several farms), and co-digestion (joint treatment of organic waste from different sources in a single farm).

Acetylastic methanogenic bacteria
The implementation of integrated organic waste management systems by geographic area (usually agricultural and industrial) also allows for the implementation of integrated organic waste management systems by geographic area, with social, economic and environmental benefits.

As can be seen in the figure, anaerobic digestion is a very complex process, both for the number of biochemical reactions that take place and for the number of microorganisms involved in them, characterized by the existence of several consecutive phases differentiated in the process of degradation of the substrate (generic term to designate, in general, the food of the microorganisms), involving 5 large populations of microorganisms.

The biochemical and microbiological studies carried out so far, divide the process of anaerobic decomposition of organic matter into four phases or stages:

1. Hydrolysis
2. Fermentative or acidogenic stage
3. Acetogenic stage
4. Methanogenic stage

The first phase is the hydrolysis of complex particles and molecules (proteins, carbohydrates and lipids) which are hydrolyzed by extracellular enzymes produced by the acidogenic or fermentative microorganisms. As a result, simpler soluble compounds (amino acids, sugars and long-chain fatty acids) are produced and metabolized by the acidogenic bacteria, giving rise mainly to short-chain fatty acids, alcohols, hydrogen, carbon dioxide and other intermediates. The short-chain fatty acids are transformed into acetic acid, hydrogen and carbon dioxide by the action of acetogenic microorganisms. Finally, methanogenic microorganisms produce methane from acetic acid, H_2 and CO_2.

In the fermentative or acidogenic stage, the fermentation of the soluble organic molecules into compounds that can be directly utilized by the methane bacteria takes place genic (acetic, formic, H_2) and more reduced organic compounds (propionic, butyric, valeric, lactic and ethanol mainly) that have to be oxidized by acetogenic bacteria in the next stage of the process. The importance of the presence of this group of bacteria lies not only in the fact that they produce the food for the subsequent groups of bacteria, but also in the fact that they eliminate any trace of dissolved oxygen from the system. This group of microorganisms is composed of facultative and obligate anaerobic bacteria, collectively called acid-forming bacteria.

In the acetogenic stage, while some fermentation products can be directly metabolized by the methanogenic organisms (H_2 and acetic), others (ethanol, volatile fatty acids and some aromatics) must be transformed into simpler products, such as acetate (CH_3COO-) and hydrogen (H_2), by the acetogenic bacteria.

In the methanogenic stage, a large group of strict anaerobic bacteria acts on the products resulting from the previous stages. The methanogenic microorganisms can be considered as the most important within the consortium of anaerobic microorganisms, since they are responsible for the formation of methane and the elimination from the medium of the products of the previous groups, being, in addition, those that give name to the general process of biomethanization.

Methanogenic microorganisms complete the anaerobic digestion process by forming methane from monocarbon substrates or substrates with two carbon atoms linked by a covalent bond: acetate, H_2/CO_2, formate, methanol and some methylamines.

In general, the speed of the process is limited by the rate of the slowest stage, which depends on the composition of each waste. For soluble substrates, the limiting stage is usually methanogenesis, and to increase the rate the strategy is to adopt designs that allow a high concentration of acetogenic and methanogenic microorganisms in the reactor. With this, systems with process times in the order of days can be achieved. For wastes in which the organic matter is in particulate form, the limiting phase is hydrolysis, an enzymatic process whose rate depends on the surface area of the particles. Usually, this limitation leads to process times in the order of two to three weeks. To increase the speed, one of the strategies is pretreatment to decrease the particle size or aid solubilization (maceration, ultrasound, heat treatment, high pressure, or combination of high pressures and temperatures).

2.3. - Biogas

Biogas is a gaseous mixture consisting mainly of methane and carbon dioxide, but also contains various impurities. The composition of the biogas depends on the digested material and the operation of the process. When the biogas has a methane content of more than 45% it is flammable. Biogas has specific properties as listed in Table 2.1.

Table 2.1.- Biogas properties

Composition	$-CH_4$ 55 a70% $-CO_2$ 25 a45% $-N_2$ 2 a 7% $-H_2$ 1 a 5% $-H_2S50$ at 5000 ppm
Properties	- **Density 0.8 to 1.2 kg m^{-3}** - **Smell: Rotten egg** - **Color: Colorless** - **Toxic**
Energy content	**6.0 - 6.5 kW h m^{-3}**
Fuel equivalent	**0.60 - 0.65 L petroleum/m^{-3} biogas**
Explosion limit	**6 - 12 % of biogas in the air**
Ignition temperature	**650 - 750°C (with the above CH4 content)**

Critical pressure	**74 - 88 atm**
Critical temperature	**-82.5°C**
Molar mass	**16,043 kg kmol^{-1}**

Due to its high methane content, it has a heating value somewhat higher than half the heating value of natural gas. A biogas with a methane content of 60% has a heating value of about 5,500 kcal/Nm^3 (6.4 Wh /Nm^3 . That is, except for the H2S content, it is an ideal fuel.

2.4. - Benefits and uses of biogas

Like natural gas, biogas has a wide variety of uses, but being derived from biomass, it is a renewable energy source. There are several benefits derived from the process of converting organic waste into biogas.

In many countries, biogas production is subsidized or has economic incentives, providing farmers with additional income. Therefore, in the agricultural sector, the implementation of anaerobic digestion technologies can provide important economic, environmental and energy benefits. Moreover, it allows for improved nutrient management, reduced greenhouse gas emissions and the capture and use of biogas. When organic waste undergoes aerobic degradation, low energy compounds such as CO2 and H2O are generated. Much of the energy is lost and released into the atmosphere. It is estimated that the energy loss of an aerobic process is approximately twenty times higher than that of an anaerobic process.

In the case of anaerobic degradation, products of metabolism with high energetic power are generated (e.g., alcohols, organic acids and methane), which serve as nutrients for other organisms (alcohols, organic acids), or are used for energy purposes by society (biogas).

Methane production is a direct result of COD reduction within the methanogenic system. Anaerobic treatment is often used to treat wastewater with high COD, for a removal efficiency of 71-97%.

For an anaerobic system, the COD (Qwmical Oxygen Demand) can be considered a conservative parameter, i.e. the sum of the input COD must be equal to the sum of the output COD: influent COD = effluent COD + biogas COD.

If we consider a biogas formed exclusively by CH4 and CO2, and taking into account that the COD of CO2 is zero, the COD eliminated in the waste would correspond to the COD obtained in the form of methane, which means 2.857 kg COD per m^3 CH4, or 0.35 m^3 of CH4 per kg of COD eliminated, at P=1 at and T=0° C or 0.38 m^3 of CH4 at P=1 at and 25° C. Taking into account the calorific value of methane, these values would correspond to approximately 3.5 kWh/kg COD.

removed, in units of primary energy. This gives anaerobic systems a clear advantage over aerobic systems for organic waste and wastewater treatment, for which the energy consumption for oxygen transfer is around 1 kWh/kg O2 consumed.

Variations from the above values may be due to possible accumulations in the reactor, to the production of other gases (H2, H2S,), or to the fact that the measured COD is not only due to oxidizable carbon.

Figure 2.3 illustrates this conservation of COD in the anaerobic digestion process for a waste with 10% of the COD non-biodegradable, where 90% of the initial COD is transformed into CH4.

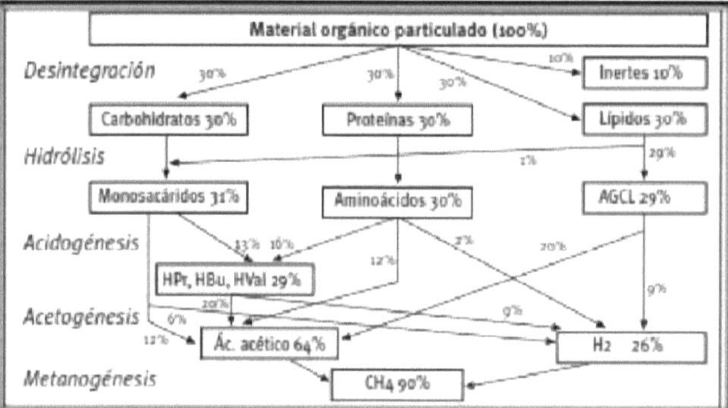

Fig. 2.3.-Chemical *Oxygen Demand (COD)* flux. *Source: Batstone et al., 2002 and GIRO. In the process of anaerobic digestion of particulate organic matter, formed by 10% of inert materials and 90% of carbohydrates, proteins and lipids, in equal parts. Propionic acid (HPr, 10%), butyric acid (HBu, 12%) and valeric acid (HVal, 7%) have been grouped together to simplify the scheme. LCFA: long chain fatty acids.*

Another important environmental benefit of biogas plants is the significant reduction of pressure on municipal solid waste. In this way, the costs of organic waste disposal are significantly reduced, and even value-added by-products (e.g. biofertilizer) are obtained. Furthermore, the anaerobic treatment of organic waste contributes to the protection of groundwater, reducing the risk of nitrate leaching.

Moreover, anaerobic digestion eliminates the problem of nuisance odors, such as, for example, the ammonia odor resulting from the accumulation of untreated excreta and urine. The promotion and implementation of collective biogas production systems for several farms and co-digestion systems for the joint treatment of organic waste from different sources in a geographical area, usually agricultural and industrial, also allows the implementation of integrated management systems for organic waste by geographical area, with social, economic and environmental benefits.

Anaerobic digestion can be carried out with one or more wastes, provided that they are liquid, contain fermentable material, and have a relatively stable composition and concentration. Co-digestion is a technological variant that can solve problems or deficiencies of one waste, if they are compensated by the characteristics of another.

Methane is a gas in the Earth's atmosphere that contributes to the greenhouse effect. The methane content in the atmosphere has doubled since the last ice age to 1.7 ml m^{-3} today. This value has remained constant in recent years. Methane contributes 20% to the anthropogenic greenhouse effect. Among the human sources of methane, more than 50% comes from livestock and up to 30% from rice cultivation.

In order to be able to compare the effect of different greenhouse gases, each is assigned a factor representing a measure of its greenhouse effect or global warming potential, compared to the CO_2 that is used as the "reference gas" (Table 2.2). The CO_2 equivalent of greenhouse gases can be calculated by multiplying the greenhouse potential in relation to the mass of the respective gas. It indicates the amount of CO_2 that produces the same greenhouse effect in 100 years, i.e. CH4 is a greenhouse gas more potent than CO_2 by a factor of 21.

Another important environmental benefit of biogas plants is the significant reduction of pressure on municipal solid waste. In this way, the costs of organic waste disposal are significantly reduced, and even value-added by-products (e.g. biofertilizer) are obtained. Furthermore, the anaerobic treatment of organic waste contributes to the protection of groundwater, reducing the risk of nitrate leaching.

Moreover, anaerobic digestion eliminates the problem of nuisance odors, such as, for example, the ammonia odor resulting from the accumulation of untreated excreta and urine. The promotion and implementation of collective biogas production systems for several farms and co-digestion systems for the joint treatment of organic waste from different sources in a geographical area, usually agricultural and industrial, also allows the implementation of integrated management systems for organic waste by geographical area, with social, economic and environmental benefits.

Anaerobic digestion can be carried out with one or more wastes, provided that they are liquid, contain fermentable material, and have a relatively stable composition and concentration. Co-digestion is a technological variant that can solve problems or deficiencies of one waste, if they are compensated by the characteristics of another.

Methane is a gas in the Earth's atmosphere that contributes to the greenhouse effect. The methane content in the atmosphere has doubled since the last ice age to 1.7 ml m^{-3} today. This value has remained constant in recent years. Methane contributes 20% to the anthropogenic greenhouse effect. Among the human sources of methane, more than 50% comes from livestock and up to 30% from rice cultivation.

In order to be able to compare the effect of different greenhouse gases, each is assigned a factor representing a measure of its greenhouse effect or global warming potential, compared to the CO_2 that is used as the "reference gas" (Table 2.2). The CO_2 equivalent of greenhouse gases can be calculated by multiplying the greenhouse potential in relation to the mass of the respective gas. It indicates the amount of CO_2 that produces the same greenhouse effect in 100 years, i.e. CH_4 is a greenhouse gas more potent than CO_2 by a factor of 21.

Table 2.2. Warming potential of greenhouse gases.

Gas	Warming potential
CO2	1
CH4	21
N2O	310
SF4	23900
CBP	9200
HFC	11700

Biogas is used as any other fuel for domestic and industrial use, the indispensable prerequisite is the availability of burners specially designed to operate with biogas.
Some of the devices in which it could be used are:

- Stoves
- Lamps
- Refrigerators
- Heaters
- Incubators
- Power Generation Engines

2.4.1. - Equipment for the generation of electricity with biogas

Using biogas as fuel in motor generators:

- Dual fuel engines
- Special engines for biogas (including gas pre-treatment)
- Gas Generators - Versatile to run on biogas and other alternative gases (when running on biogas, 10% of engine power is lost, which leads to the same level of losses in electricity generation).

Diesel engines adapted to run on biogas

Generator with adapted diesel engine

Generator with biogas engine

Fig.2.4.- Biogas motors

Gas generators with Diesel engine:
- In the case of diesel engines, biogas can replace up to 80% of diesel fuel.
- The low ignition capacity of biogas does not allow to replace all the diesel fuel in this type of engines that do not have a spark plug for combustion.
- The gas is sucked together with the combustion air into the cylinder.

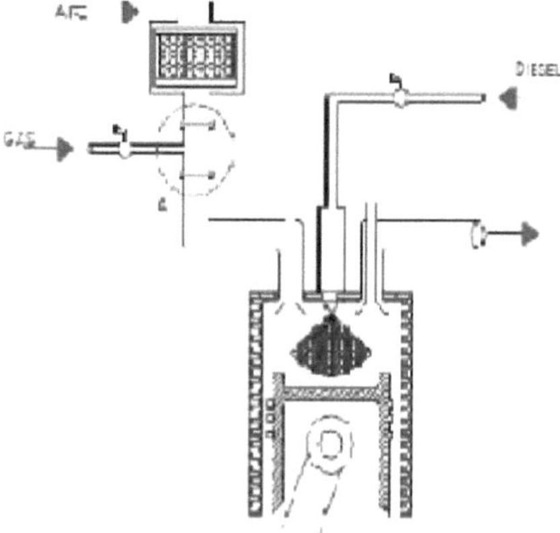

Fig.2.5.- Device for feeding biogas to the engine

Devices to adapt engines for biogas utilization:
1. Filter to capture hydrogen sulfide in biogas
2. Air-Biogas Mixer

- There is no loss of pressure during mixing, increasing the power to maximum efficiency.
- Even with changes in the gas mass flow rate, the air-gas ratio remains constant.
- With the combustion chamber temperature, the control system regulates the engine exhaust emission by adjusting the air-gas ratio accordingly.

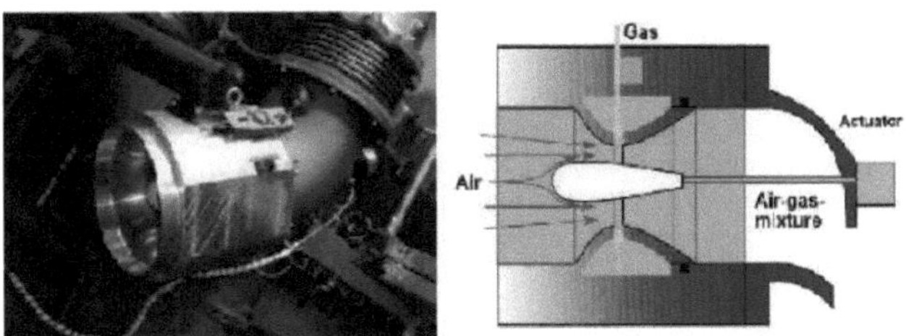
Fig.2.6.-Devices for feeding biogas to the engine

In the energy market, the biogas plant competes with wood, propane gas and electricity, energy sources commonly used for cooking; with kerosene, candles and electricity for lighting, especially in places where service is deficient or non-existent; with propane gas and electricity for refrigeration; and with gasoline or diesel as motor fuel.

One m^3 of biogas can replace 0.46 kg of propane gas, 0.7 liters of gasoline, 0.6 liters of diesel or 2 kg of wood, which greatly prevents the destruction of forests. It has been calculated that 1 m^3 of biogas used for cooking avoids the deforestation of 0.335 ha of forest with an average of 10 years of tree life (Sasse, 1989).

One cubic meter of fully combusted biogas is sufficient for:
Generate 1.25 kW/ h of electricity
Generate 6 hours of light equivalent to a 60-watt light bulb
Run a refrigerator with a capacity of 1 m^3 for 1 hour.
Run an incubator with a capacity of 1 m^3 for 30 minutes.
   Running a 1 HP motor for 2 hours

In the fertilizer market, the biogas plant competes with fresh manure and chemical fertilizers, since it allows a saving in the amount of other conventional fertilizers without decreasing productivity and also presents an increase in productivity when compared to unfertilized soils. In the waste treatment market, biogas production can prevail over traditional aerobic treatment systems, which are much more expensive and complex.

Studies carried out in Cuba have shown that the use of liquid effluent is economically more profitable than biogas itself (Carballal, 1998).

Table 2.3. Minimum duration of 1m³ of biogas for different artifacts

Feeding	Consumption Kcal/h	1 m³ biogas (minimum duration)
1 burner stove	660 - 742.5	7.4 h
84.95 L refrigerator	550 -600	8.3 h
Lamp	478 -528	10.4 h
110 L thermos tank	1375 - 1650	3.3 h
600 cal stove	3355 -4400	1.25 h
Engine (hp/h)	2750 - 4400	1.25 h
Electricity Generation	6.4 kW/h	2h

2.4.2. - Utilization of biogas residues

Residues are by-products of biogas. The use of biogas residues can have more economic benefits than the use of biogas itself, and in many countries, including China, India and the Philippines, much research is being done on this option.

Residue is a type of organic fertilizer and possesses the characteristics of traditional fertilizers and other advantages as a result of anaerobic fermentation.

In anaerobic decomposition, most of the nutrients are retained, except for some elements such as carbon, hydro and oxide which are transformed into CH4 and co2. The soluble nutrients remain in the residue as a chemical residue, and at the same time some solid substances, both organic and inorganic, absorb nutrients during the process. Therefore, biogas residues are more nutritious than traditional fertilizers.

In addition to nutrients N, P, K, biogas fertilizer contains other substances abundant for plants; it has a higher concentration of humic acids, cellulose, hemicellulose than compost. That is why it fertilizes the soil more.

Table 2.4. Chemical components

Chemical components of fertilizers	N (%)	P(%)	K(%)
Biogas Residues	1,45	1,10	1,10
Dry biogas residues	1,60	1,40	1,20
Estiercol	1,22	0,62	0,80
Compost	1,30	1,00	1,00

Research results and practical experiences have shown that biogas residues can reduce insects and weeds by active biological substances such as gibberilin, acetic acid, hormones..., which do not exist (or in negligible quantity) in original materials, but abound in residues.
When organic substances are fermented, a part of nutrients are absorbed by bacteria for gas production, and another part is transformed into amino acids at a rate of 230% compared with original materials. Also, an amount of vitamin B12 is composed during fermentation. According to one investigation, the amount of B12 in one cubic meter of dry waste is about 3000 kg when the corresponding figures in fish powder and bone are 200 and 100 kg respectively.

The utilization of biogas residues as fertilizer for plants or animal feed will be economically beneficial.

The residues are divided into two types: the Kquids which contain dissolved substances and the condensates which are sediments at the bottom of the digester. Biogas residues can be used directly without compounding with chemical fertilizers:

Liquid residues comprise dissolved nutrients that can be easily absorbed by plants. It is used in the same way as traditional fertilizers (additional fertilization).

Condensed residues have a high concentration of nutrients, more abundant in organic substances and humic acids. Moreover, they have both rapid and prolonged effects and are therefore suitable for basic fertilization (before sowing seeds). If they are used for land with scarce water, these residues should cover the surface of the field by a layer of 0 to 12 centimeters; and in the case of land with abundant water, it is necessary to spread the material over the entire surface and then lift the soil well so that they are mixed together.

If the fermentation materials are mostly pig droppings, you should add more water to the waste before using it. In this case, the liquid waste, when freshly removed from the decomposition pond, will continue to ferment and, if spread, will absorb oxygen from the plants, causing damage to their roots until they dry out. For best effects, you should keep residues, after removing them from the digester, in another tank for a few days for the residuals and two weeks for the condensates.

The residues can be mixed with chemical fertilizers. This combination will balance the different nutrient needs between soil and plants. The addition of protein sulphate and proternate carbonate to biogas residues accelerates nutrient production, which results in delayed nitrogen depletion and increases the ratio of chemical fertilizer. The highly nutritious biogas residues are able to boost the growth of plants and microorganisms, which decreases the need for chemical fertilizers and prevents the destruction of soil structure.

Compost is an organic fertilizer whose main components are botanical materials. Compost includes a lot of cellulose and lacks nitrogen, so the addition of enzymes, microorganisms or auxiliary substances is necessary for rapid decomposition. Biogas residues, among other nitrogen-rich compounds, can be used as a source of enzymes.
Drying is a simple but efficient method for preservation and transportation. However, this option will cause a notable loss of nutrients, especially nitrogen.

Biogas residues can be used in a similar form as manure. Dried residues can be used as dried animal excrement.

Biogas residues contain many substances beneficial to animals such as the elements calcium, phosphorus, nitro, copper, zinc and iron; many types of proteins, celluloses and amino acids, including nine acids indispensable to animals. The enzyme also increases the effectiveness of feed use. Fermented waste from the excrement of large livestock decreases the deficiency of vitamins B1 and B2.

Liquid pig waste can be used directly or with diluted traditional feeds. Pigs over 20 kg are adaptable to feed with biogas residues and the volume of residues used increases gradually with the weight of the pig. In case the pig has diarrhea, the percentage of residues in the feed should be reduced.

     Requirements to be met:

- The biogas residues for this use should be the surface part in a digester of normal conditions (decomposition tank of at least one month of operation).
- Materials containing dead animal carcasses, insecticides, herbicides or other toxins must not be introduced into the digester.
- Do not feed pigs weighing less than 20 kg and mother pigs with biogas residues.
- Deworming animals before supplying them with biogas residues.

Biogas residues are used to feed fish both in ponds and in rice fields. In case of fish in ponds:

- . The use of biogas residues depends on the rate of omrnivorous fish present in the environment. If this rate is lower than 30 percent, you can use biogas residues as the main feed, and this rate is higher than 40 percent, you have to mix residues with other feeds.

- . In general, condensed wastes are used as a basic feed and the residuals are used as an additional feed. It is necessary to spread the waste in the air for a while before introducing it into the lagoon.

In addition to adjusting the volume of waste introduced into the lagoon according to the temperature, it is necessary to enrich the oxygen in the water. One of the simple methods to determine the oxygen level in the lagoon is to measure the time the fish leave the surface of the water: the longer the fish leave, the less oxygen there is in the lagoon.

In case of fish in paddy fields, creating a network of parallel horizontal and vertical ditches, each 30 cm deep and 30 cm wide. Around the paddy dig larger ditches (50 cm deep and 60 cm wide) for fish housing.

One week after transplanting the rice, they start to stock the rice field with fish, and in the following period, every eight days, biogas residues are spread once in the major irrigation ditches, but not in the places where fish are found. Contact between fish and insecticides should be avoided when using insecticides to protect the crop.

2.4.3. - Biogas purification

Biogas purification is nothing more than the removal of carbon dioxide and hydrogen sulfide. The carbon dioxide is removed to increase the fuel value of the biogas. CO_2 is a colorless, odorless gas, weighs 1.5 times the weight of air and is non-combustible. A high percentage of this gas in the compound will reduce the quality of the biogas.

Hydrogen sulfide is removed to reduce the corrosion effect on metals in contact with the biogas (Hesse, 1983). Hydrogen sulfide (H_2S) is another colorless gas, which is also non-combustible and occupies a very small percentage of the compound. However, it produces a foul odor, which is typical of biogas.

For rural communities it is more practical not to deal with carbon dioxide removal. Farmers generally prefer a less efficient gas to having to spend time on carbon dioxide control, so on small farms this work is considered unnecessary. For large biogas plants and other specific plants where the technical aspects are less onerous, there are economic justifications for purification.

The lower the humidity, the greater the ease of combustion. Water can be removed by passing the gas through quicklime, although this affects the percentage of carbon dioxide.

The presence of carbon dioxide in the gas presents the most serious aspect: it reduces the calorific value of the fuel, and furthermore, it increases the storage capacity, as well as increases the pressure of the storage tanks. This also causes low effectiveness at the time of gas combustion, as it requires some of the heat produced to raise its ignition temperature. In spite of this, the absorption operation is made simple by passing the gas through lime water.

The use of this absorbent is no longer practical and affordable when working on a large scale, in this case, substances such as diethyl amine, triethyl amine, calcium hydroxide, potassium carbonate and potassium hydroxide are used. A cost-benefit analysis of the gain in calorific value versus reagent, water and energy consumption will undoubtedly lead to the conclusion that it is more advantageous to use CO2 gas.

Hydrogen increases the calorific value of the gas, so it is not necessary to eliminate it.
On the other hand, hydrogen sulfide is present in small, almost imperceptible quantities when the digestion cycle is extended for more than thirty days. This component
affects when the gas is used in the operation of machinery, as it helps the deterioration of the metal; if the use of the gas is only for combustion, the removal of hydrogen sulfide is not important.

The simplest and most efficient chemical method of carbon dioxide removal is its absorption in lime water. The method needs much attention because the lime water is exhausted and needs frequent replacement, resulting in its frequent preparation if it is not commercially available. Lime water can be replaced by an aqueous solution of ethanolamine, which absorbs carbon dioxide (and also hydrogen sulfide). Although this process is expensive to make routine in the purification of biogas due to the periodic heating to which this substance has to be subjected for its regeneration.

Another alternative is to use another strongly alkaline waste as an absorption medium for these gases such as microalgae culture effluents. The liquid effluent from the digester is discharged directly into a large tank to produce Spirulina algae. The algae is filtered to be used as pig or duck feed or as an additive and the waste water that has a pH value of 10 or more is stored in a cylindrical tank. This water is then counter-flowed through the biogas. The water that remains as a result of this reaction contains hydrogen carbonate which is reused in the algae cultivation.

Carbon dioxide is quite soluble even in neutral water (878 cc / liter at 20° C) under atmospheric pressure, so washing with ordinary water is perhaps the simplest method of removing impurities. CO2 is soluble in water while methane is not. At high pressure, the solubility of CO2 increases proportionally allowing the concentration of methane in the biogas to increase (Lau-Wong, 1986).

For the preservation of biogas operated equipment, especially in engines, the hydrogen sulfide (H2S) contained in the gas must be extracted. Several systems are used to achieve this purification:

a) Iron oxide (FeO2) filters. For this, iron chips can be used, which can be regenerated by exposure to open air. The air must be carefully injected into the filter and can be done with aquarium pumps.

b) Addition of FeO2 to the substrate. By adding 500 g per 4000 l of substrate, the H2S content goes from 0.2% to 0.07%. This quantity should be supplied daily.

c) Exploitation of water condensation. When large amounts of water vapor are condensed from the biogas, large amounts of H2S are absorbed to the H itself, reaching removals of 30 to 40% of the acid. This method is widely used in cold climates.

d) By addition of air. Air can be injected in a proportion of 3% to 5% directly into the digester or gas storage site, so that H2S is decomposed into water and elemental sulfur. This sulfur can be added to the liquid manure and is beneficial to the soil. The air supply must be

controlled, so as not to create an explosive mixture.

In addition to the traditional methods of desulfurization with iron filings, there is a procedure based on the addition of air to 1.5% of the volume of biogas produced (Henning, 1986). This method ensures a decrease in H2S content of approximately 120 ppm or 0.012% by volume of biogas.

2.5.- Factors influencing anaerobic digestion

Like any biological process, anaerobic digestion must be controlled, as there are several factors that have a considerable influence on its success or failure. An imbalance in any of these factors can lead to the rupture of the equilibrium between the microbial communities and consequently to the non-functioning of the system, the non-production of biogas.

Therefore, it is important to examine some of the important factors that govern the methanogenic process. Microorganisms, especially methanogenic microorganisms, are highly susceptible to changes in environmental conditions. Many researchers evaluate the performance of an anaerobic system based on the rate of methane production, because methanogenesis is considered a limiting step in the process. Because of this, anaerobic biotechnology requires careful monitoring of environmental conditions. Some of these environmental conditions are: temperature (mesophilic or thermophilic), type of feedstock, nutrient and trace mineral concentration, pH (generally close to neutral), toxicity and optimum redox conditions. These conditions are discussed below:

Temperature

This is a very important variable because as the temperature increases so does the metabolic activity of the bacteria, requiring less retention time for the fermentation process to be completed.

Anaerobic processes, like many other biological systems, are strongly temperature dependent. The reaction rate of biological processes depends on the growth rate of the microorganisms involved, which, in turn, depends on temperature. As the temperature increases, the growth rate of the microorganisms increases and the digestion process is accelerated, resulting in higher biogas yields.

The operating temperature of the digester is considered one of the main design parameters, due to the great influence of this factor on the rate of anaerobic digestion. Sudden temperature variations in the digester can trigger the destabilization of the process. Therefore, to guarantee a homogeneous temperature in the digester, an adequate agitation system and a temperature controller are essential.

There are three temperature ranges in which anaerobic microorganisms can work: > Psicrophilic (below 25°C), > Mesophilic (between 25 and 45°C)

> Thermophiles (between 45 and 65°C),

The maximum specific growth rate (iimax) is higher as the temperature range increases. Within each temperature range, there is an interval for which this parameter becomes maximum, thus determining the optimum working temperature in each of the possible ranges of operation (Figure 2.6).

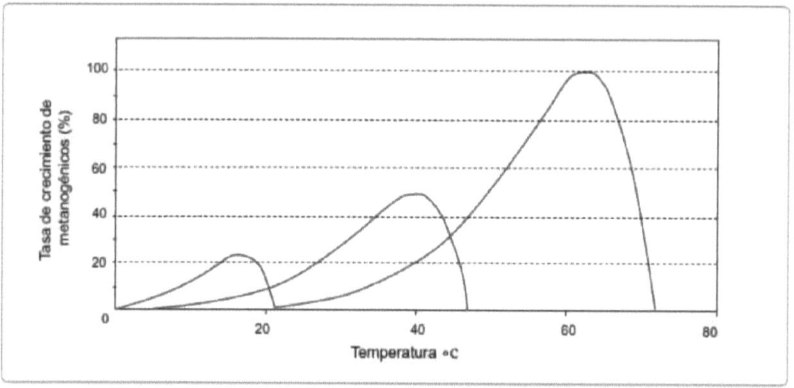

Fig. 2.6.- Relative growth rate of microorganisms. Source: Speece (1996)

The higher the temperature, the more agility is obtained in the development of the process, allowing the possibility of using smaller dimensions in the reactor, however, when working at very high temperatures the process may not be profitable, so it is common for digesters to operate in a mesophilic range. Table 2 shows the maximum, minimum and optimum values at which anaerobic fermentation can operate.

Table. 2.5.-Temperature ranges for anaerobic fermentation

	RANGES, 0C			
Fermentation	Minimo	Optimum	Maximo	Fermentation time
Sicrofflica	4-10	15-18	25-30	More about 100 days
Mesofflica	15-20	28-33	35-45	From 30 - 60 days
Termofflico	25-45	50-60	75-80	10-15 days

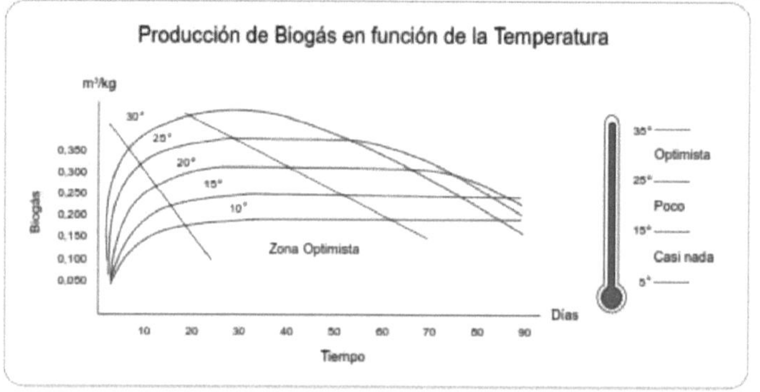

Fig.2.7.- Biogas production as a function of temperature. Source: Varnero, 1991

Carbon/nitrogen ratio

Virtually all organic matter is capable of producing biogas when subjected to anaerobic fermentation. The quality and quantity of biogas produced will depend on the composition and nature of the waste used. Nutrient levels must be above the optimum concentration for methanobacteria, as they are severely inhibited by lack of nutrients.

Carbon and nitrogen are the main sources of nutrition for methanogenic bacteria. Carbon is the source of energy and nitrogen is used for the formation of new cells. These bacteria consume 30 times more carbon than nitrogen.

Fermentation materials are mostly composed of carbon (C) and nitrogen (N). If the nitrogen content is too high, bacterial reproduction is inhibited due to high alkalinity. The ideal is a C/N ratio of 20:1 to 30:1; the decomposition of materials with high carbon content, higher than 35:1, occurs more slowly, because the multiplication and development of bacteria is low, due to the lack of nitrogen, but the period of biogas production is longer, lower C/N ratios, for example 8:1, inhibit bacterial activity due to excessive ammonium content. The ammonium concentration in the fermentation material should be less than 2000 mg/L. Table 2.6 shows the C/N ratio for some substrates.

Table 2.6.- C/N ratio of some residues

Materials	% C	% N	C/N
Animal waste			
Cattle	30	1.30	25:1
Equines	40	0.80	50:1
Sheep	35	1.00	35:1
Swine	25	1.50	16:1
Goats	40	1.00	40:1
Rabbits	35	1.50	23:1
Hens	35	1.50	23:1
Ducks	38	0.80	47:1
turkeys	35	0.70	50:1
Excretashuman	2.5	0.85	3:1
Vegetable waste Straw			
wheat	46	0.53	87:1
Barley straw	58	0.64	90:1
Straw rice Straw	42	0.63	67:1
oats	29	0.53	55:1
Corn stubble	40	0.75	53:1
Legumes	38	1.50	28:1
Vegetables	30	1.80	17:1
Tnbcrnlos Leaves	30	1.50	20:1
dry	41	1.00	41:1
Sawdust	44	0.06	730:1

Based on the carbon and nitrogen content of each of the raw materials, the C/N ratio of the mixture can be calculated by applying the following formula (1):1.

$$K = \frac{C_1 \times Q_1 + C_2 \times Q_2 + \cdots \cdots C_n \times Q_n}{N_1 \times Q_1 + N_2 \times Q_2 + \cdots \cdots N_n \times Q_n} \qquad (1)$$

K = C/N of the raw material mixture.
C = % of organic carbon contained in each raw material.
N = % of organic nitrogen contained in each raw material.
Q = Fresh weight of each material, expressed in kilograms or tons.

pH control

It is of vital importance to the system, since a decrease in pH can result in inhibiting the growth of methanogenic bacteria, resulting in decreased methane production, increased carbon dioxide content and unpleasant odors due to increased hydrogen sulfide content.

In general, the pH remains fairly stable in spite of acid production by the bacteria, since buffers are generated in the fermentation medium to ensure a suitable pH range. Furthermore, the rate of acid formation depends on the rate of conversion to biogas. It is generally accepted that the optimum pH values range from 5.5 to 8.0, however, in the two-stage system the recommended pH depends on the anaerobic phase, as shown in Table 2.7.

Table .2.7.- Optimal pH for biogas production.

pH value	Hydrophilic stage	Methanogenic stage
Typical value	5.0-6.0	6.5-7.5
Optimum value	5.5-5.7	6.8-7.2

Suitable bacteria

There must be an optimal ratio of both methanogenic and non-methanogenic bacterial populations, which is guaranteed by a previous inoculum, which develops sufficient buffering substances to maintain the desired pH values and which almost completely cover the high demands of anaerobic conditions by the methanogenic bacteria.

System inhibition

All organic matter is composed of water and a solid fraction called **total solids** (TS). The percentage of total solids contained in the digester loading mix is an important factor to consider to ensure that the process is carried out satisfactorily. The most favorable percentages of total solids in the fermentation medium should be between 5 and 12% for semi-continuous digesters, since values of 15% and above tend to inhibit the process for batch digesters, which have between 40 to 60% total solids. The mobility of methanogenic bacteria within the substrate becomes increasingly limited as the solids content increases and therefore efficiency and gas production may be affected.

Heavy metals, antibiotics, high concentrations of ammonia, mineral salts and some substances such as detergents and pesticides are products that inhibit the biogas production process, there should be no conditions in the system that facilitate the entry of oxygen or the presence of compounds that oxidize and release oxygen, such as nitrates, because methanogenic bacteria require strict anaerobic conditions.

CHAPTER 3.- DESIGN AND CONSTRUCTION OF BIOGAS PLANTS

The technology of biodigesters, anaerobic digesters, anaerobic reactors or biogas systems has become very popular in the last decade. The strong dependence on oil-based energy sources, the concern of the civilian population and the State about global warming, as well as the pollution of water sources, have been the pillars of its notoriety. The biodigester, as it is popularly known, offers simple solutions to the problems mentioned above. On the one hand, it provides a selected system, par excellence, for the treatment of wastewater (green water), prior to its use as fertilizer in the paddocks. On the other hand, during the decomposition process, it generates a gas with high methane content, which allows its use as an energy source.

3.1. - Concepts

A biodigester is, in general terms, a hermetic compartment in which organic matter is fermented in the absence of oxygen. As a result of this process, a combustible gas is obtained, which is approximately 66% methane and 33% carbon dioxide.

The material resulting from biodigestion, or effluent, can be directly used as fertilizer and soil conditioner, as nutrients such as nitrogen become more available, while others such as phosphorus and potassium are not affected in their content and availability.

In order to calculate the size of a biogas biodigester, a number of characteristic concepts are used. For a simple biogas plant they are the following:

Sd - Daily amount of fermentation sludge (influent or substrate).

TR - Retention time.

Gd - Specific gas production per day depending on retention time and fermentation material.

DM, - the dry mass, i.e. the percentage of water that varies in each natural fermentation material. For this reason, in more accurate research work, we work with the solid part or dry mass of the fermentation material.

MOS, - dry organic mass, for the fermentation process only the organic or volatile components of the fermentation material are important. Therefore, only the organic part of the dry mass is processed.

Md - Digester load, indicates how much organic material is fed daily or how much material must be fermented per day. The digester load is calculated in kilograms of organic mass per cubic meter of digester per day (kg. MOS/m3 / d).

3.2. - Characteristics of the fermentation material

In the following tables 3.1 and 3.2, information is given on some characteristics of the fermentation material or substrate, according to the existing experience in the world and as results of research carried out in this field.

Table .3.1.- Fermentation Material

Origin	Amount of daily excreta Md (kg)	% of excreta per live weight	% urine per live weight	% mater ferme fresh MS	of ial de ntacion MOS	Rate C/N	Biogas m³ /kg excreta Y	Dissolution rate Nd Exc:Water

Cattle	15-20	5	4	16 20	13	24 25	0.04	1:1
Buffaloes	18-25	5	4	18 25	12	18 25	0.04	1:1
Pigs	1.2-4	2.5	3	17	14	12 13	0.07	1:1-1.3
Sheep	1-2			30	20	25 30	0.05	1:2-1:3
Equines	10-15			25	15	2425	0.04	1:2-1:3
Birds	0.02-0.08			25	16	5-16	0.06	1:3
Humans	0.18-0.5			20	15	2.9 10	0.07	1:2-1:3

C/N= Carbon to Nitrogen Ratio

Table .3.2.- Fermentation Material

Origin	Daily quantity	% of ma fermentation MS	iterial of fresh m MOS	C/N rate	Biogas (liter/kg of material
Fresh hyacinth	25 kg/m²	7	5	12-25	0.3-0.4
Dry straw		80-85		48-117	1.5-2.0
Slaughterhouse waste		15-20		(0.34-0.71m3/kg DM	
Distillate residual	(15 m3 biogas/m3 of waste)				

Yeast residual	(4 m3 biogas/m3 of waste)
Coffee liquid	(5 m3 biogas/m3 of waste)
Coffee husk	(0.1 m3 biogas/m3 of waste) (0.4 m3 biogas/25 kg dry matter)

3.3. - Structure of biogas plants

In the structure of a biodigester we can first consider three zones:

Biogas Zone: This is formed by a dome at the top where the biogas accumulates. Depending on the type of reactor to be built, it can be a fixed or floating dome.

Digestion zone: This is represented by a cylinder in the intermediate part where the process of anaerobic digestion of substrates or organic matter is carried out.

Sludge zone: Consists of a cone at the bottom where the sludge is deposited.

According to the zones defined above the biodigester has 5 main components: (I) inlet pipe; (II) digester (decomposition pond); (III) outlet pipe; (IV) pressure regulating tank (fixed cover type); (V) gas collection cover (can be fixed or floating cover type), there must also be a storage place or pit for organic matter.

1. *Inlet pipe:* with the task of conveying material to the decomposition pond. This pipe has cylindrical shape, made of iron, concrete or plastic with the minimum diameter of 150mm. One end of the pipe is attached to the bottom of the inlet tank and another point is attached to the digester, with an angle deviation of 30 degrees from the vertical direction. This position ensures easy entry of material into the tank, also has the function of moving the decomposed material and giving better contact to the mass of bacteria.

2. *Digester (decomposition pond):* is the most important component of the equipment. It contains the decomposed liquid and the place where the fermentation that generates biogas takes place. The inlet and outlet pipes are installed symmetrically on the two opposite sides of the pond.

3. *Outlet pipe:* it has the same shape and structure as the inlet pipe, but the diameter can be smaller because the waste comes out in Kquid mode and can be easily removed. This pipe is tied to the pond at a 40 degree angle inclination to the vertical direction.

4. *Pressure regulating tank:* semi-spherical shape, pressure regulating function in the decomposition tank. In addition, it has the function of containing liquid after decomposition and is a safety valve for the pond. An overflow pipe installed near the pond mouth has the function of preventing overloading when too much gas is produced. When the liquid in the pressure regulating tank reaches a certain level, it flows out through this tube.

5. *Gas collection cover: It* can be made of steel or solid plastic or construction material depending on the type of biodigester chosen, with the gas pipe inside the tank and crossing the cover to conduct the biogas out of the digester.

3.4. -Qualification of biogas plants

Among the most widely used and marketed groups of anaerobic reactors for the treatment of first generation solids, the most common are the Hindu and Chinese types (Montalvo and Guerrero, 2003). These reactors are not of great efficiency from the point of view of biomass biodegradation, but due to their constructive and operational simplicity, their start-up makes them more appropriate to obtain biogas mainly from agricultural residues (Ringkamp and Col.1988; Sang-shi and Xi-Chun, 1988).

Floating or Hindu-type biogas plants

This biogas equipment was developed by the Khadi and Rural Industries Commission of India (KVIC). The gas container is a box-shaped lid cap directly over the liquid or water seal around the mouth of the pond (Fig. 3.1). The gas produced in the pond is collected and contained in the lid and floats the lid, the more gas it contains, the more the lid floats. The weight of the lid creates a pressure on the gas contained inside, when gas is removed for use, the lid sinks down. The introduction of new material to decompose through the inlet tube creates a pressure that pushes the residue or already decomposed material out through the outlet tube.

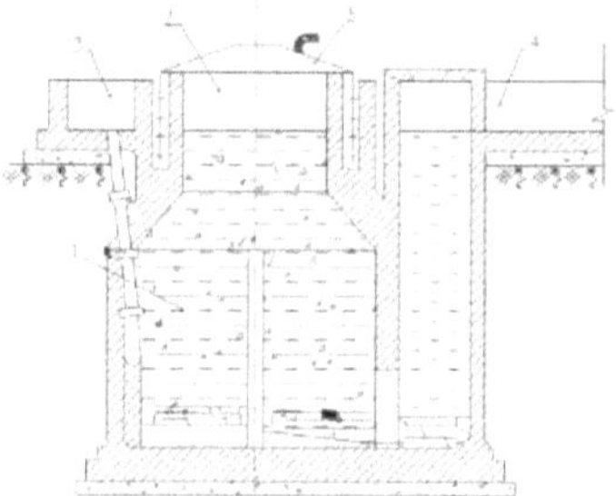

Fig. 3.1.- Floating lid biogas plant with water seal
(1- Digester; 2-gas containment cover; 3- substrate inlet; 4- waste outlet)

The cover of this equipment is made of iron or concrete with iron netting with the quality already checked before leaving the workshop. The weight of the cover has a great influence on the pressure inside the tank, which is a factor that should be taken into account during the design process. The cover made of iron is very expensive (it takes 30-40% of the total cost of the work). The cover has direct contact with the air and is affected by the ambient temperature, in the winter, the low temperature will greatly affect the gas yield. The water seal protects the tank from rain water ingress and avoids contact between liquid and outside air which causes a reduction of anaerobic and thus productivity.

Fixed cover

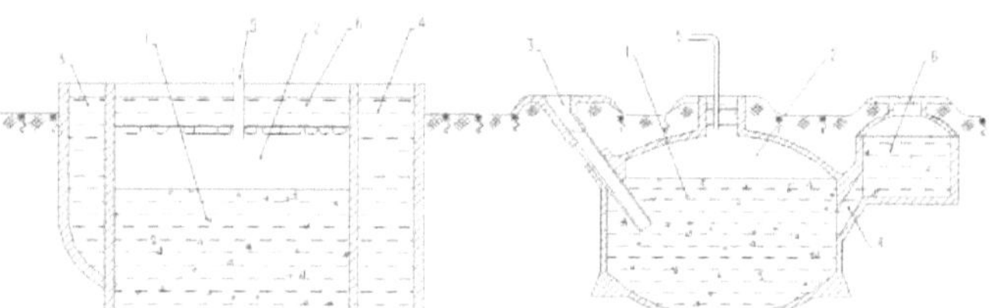

Fig.3.2.- Biogas plant with fixed cover from China

Fixed cover equipment is constructed of brick, cement, sand and has the cheapest price in comparison with the iron cover equipment. Also by the application of not so advanced techniques, the beneficiary can build on his own. The fixed cover plant is built under the ground and does not occupy space, maintains stable temperature, but in the construction it is necessary to ensure that no air enters. The following figures show some construction variants of fixed cover biodigesters.

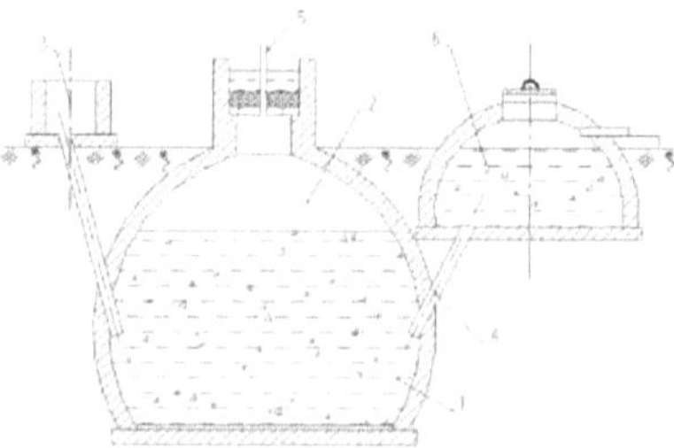

Fig. 3.3- Plant with fixed cover in spherical shape

The design can also be of the type shown in the figure below:

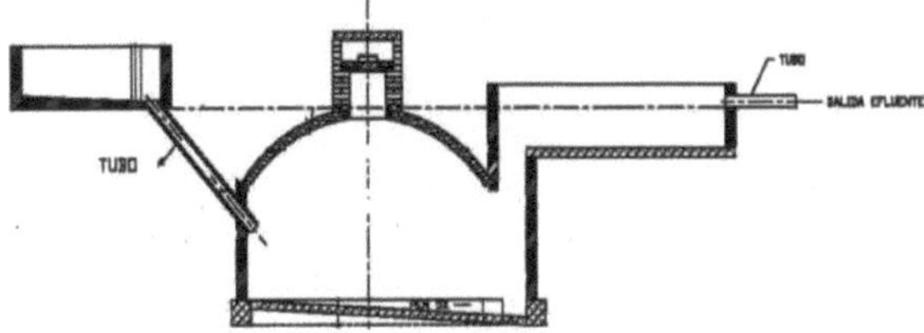

Fig. .3.4.- Fixed dome digester type

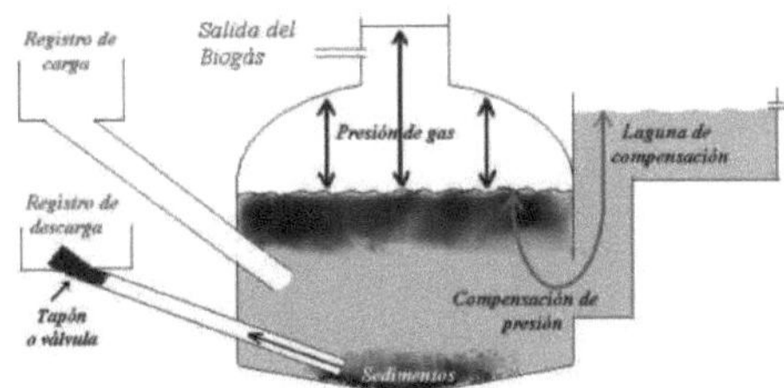

Fig. 3.5.- Fixed Dome Biodigester with sludge extraction.

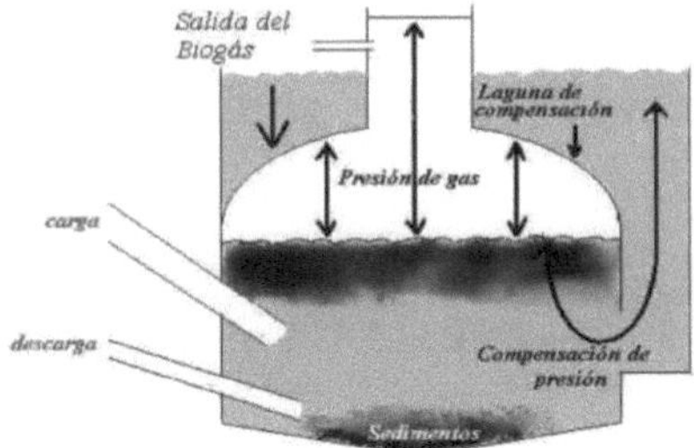

Fig.3.6.- Biodigester with compensation lagoon on top of the dome

These biodigesters shown in Figure 3.6 (Valia, 2005), according to their designer, have some advantages over others of their type. These are:

- Occupies less construction area.
- Minimizes excavation volume.
- Saves earth movement for biodigester backfilling.
- Minimizes execution time.
- The lagoon on top of the cupola favors its waterproofing.
- The biofertilizer output is realized by taking advantage of the hydrostatic loads.
- The use of the drying bed facilitates the handling of the compost.

The use of ponds, lagoons and wetlands favor the better treatment of the biodigester effluent, facilitate its use in crop irrigation which implies a considerable saving of drinking water, some pictures are shown below:

Fig.3.7.- Photo of biodigester with compensation lagoon on top of dome

Fig.3.8.-Photo of fixed dome digesters in the community of Magueyal (Palacios-Recio, 2006).

Nylon bag biodigesters

The lower part of the bag (75% volume) is filled with the fermentation mass, while the upper part of the bag (25%) stores the gas. The gas is contained in the bag due to the elastic character of the bag and the pressure regulating tank is not necessary, but it needs weight on the bag to create pressure. This equipment is easy to assemble, inexpensive, but does not serve for a long time, as shown in Figure 3.9.

Fig.3.9.- Nylon bag biogas plant

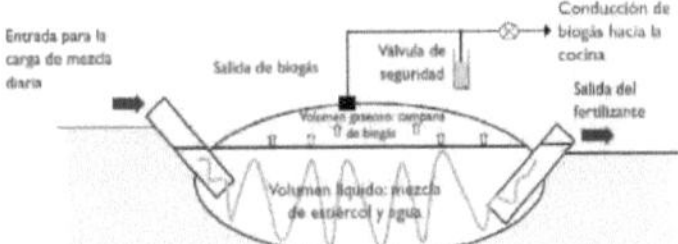

Fig.3.10.- Diagram of the operation of a nylon bag biodigester.

3.5.   - Biodigester Design Calculations

There is no generalized optimal biodigester, due to the characteristics of the productive process that conceives it, since they depend on the:

a)   User's energy demand.
b)   Amount of biomass available.
c)   Average temperature of the site.
d)   Specific gas production according to available biomass.

The knowledge of these factors allows to dimension the required volume of the biodigester, the volume of the gas storage space and the volume of the compensation tank, according to the problem to be solved. There are several references that provide minimum or starting information on the above mentioned factors.

The size of the digester (*Digester Volume*) is determined on the basis of the selected retention time (*TR*) and the amount of input substrate (*Sd*).

$$V = Vd + Vg + Vc \tag{2}$$

V=Digester volume (m)³

Vd=Volume of the substrate containment (influent organic matter mixtureand

water) Vg=Volume of biogas per day, m3 day

Vc=Volume of dead space (5-6% of Vd+Vg)

$$Vd = Sd \text{ x } TR \tag{3}$$

Sd= (1 +Nd) x Md -Daily amount of fermentation silt, liters/day

Nd=Dissolution rate, (liters of water per kg of excreta)

Md=Mass of organic matter (excreta), kg day

Md= No of animals x P_{vp} / P_{ve} x kg of daily excreta x T_e/24

P_{vp} =Average live weight of the animal population for the design

P_{ve}= Equivalent live weight.

(For pigs P_{ve}=50kg; for cattle Pve=350kg; for cattle Pve=350kg.

For pigs and cattle the volatile suspended solids are estimated at 20%, it is recommended that the organic load does not exceed 2kgm 3 day, so it is convenient to perform this check at the end.

MS / Vd < 2

TR=Retention time, days

Vg=Mdx Y, m3 day

To the product of the daily amount of fermentation sludge by the retention time, 20% of the volume is added for gas storage.

3.6. - Construction specifications of a fixed dome biodigester

The biodigesters have to be designed according to their purpose, arrangement and the temperature for the required work, therefore, a series of steps should be taken into account that are recommended to achieve success in its construction, these are:

1. **Correct selection of the construction site, which should take into consideration the following recommendations:**

   a) To guarantee the surface of the equipment construction plan with the designed dimension. Save the plane surface and does not affect the surrounding works.
   b) Away from low-lying areas, lakes, marshes... to avoid water accumulation for a long age for the construction site.
   c) Avoid weak soil area, it complicates the treatment of the base.
   d) Avoid construction near large plants, the roots will damage the construction site.
   e) Near inlet material supply, if the plant combines with human waste treatment, it should connect directly with the digester for proper

   treatment.
   f) Near gas consumption place to save pipes, avoid gas leakage in case of broken pipe.
   g) Fence of liquid waste container for use as fertilizer.
   h) Equipment should be installed in a place with sunshine, little air to maintain high temperature, creates favorable conditions for fermentation.
   i) Away from the 10meter water tank works.
   j) Away from subway works such as cables, pipes...

Before the installation of the plant, it is necessary to make a preliminary analysis or study the documents on the geographical condition of the site in order to have adequate measures of structure, construction, preparation of materials, appropriate design. The evaluations are:

   a) Evaluation of site soil condition: soil type, similarity between soil layers, it is necessary to perform analysis after completion of excavation for a more accurate result.
   b) Determine the groundwater level in the depth frame of the equipment to better measure the treatment or installation: build a ditch around to collect water, water collection well, build a suitable type of base...etc.

2. **Correct preparation of construction materials.**

After having the site design, calculate the exact amount of earth to be excavated, prepare the plan and concentrate the construction materials: sand, cement, bricks, prepare the plan for the purchase of these materials. To ensure the quality of the work, the building materials must meet the following technical requirements:

   a) *Brick*: due to the subway nature of the work, with humidity, the brick should be solid with better quality (size 65mm x 110mm x 220mm), square, clean surface.
   b) *Sand*: Yellow sand, diameter less than 3mm, clean, separated from debris. The foreign element in sand should be less than 6%. In case of non-compliant quality, it is necessary to scrub the sand before use.
   c) *Cement*: it must have the capacity to withstand pressure of $P >$ or $= 300kg/cm3$.
   d) *Gravel, stone*: these are materials necessary for concreting the base of the decomposition pond and regulation tank, the general requirement is that these materials be clean, without mixed elements.
   e) *Iron*: The types of iron used must be identical to those of the design, not rusted.
   f) *Mortar:* can be cement or a mixture of cement and lime, cement creates a solid character and avoids absorbing, but cement is brittle. Lime makes the mortar soft, and sticks better. You should use the mortar from the same factory as the brick factory, in the mortar you

should use yellow sand with cement P> or = 300kg/cm3, the ratio of cement to sand is 1:4. Mortar mixture of cement: lime: sand is

1:0,5:5.

g) *Pipes*: plant inlet and outlet connection pipe has the diameter of 100cm or more. Zinc coated iron pipe, PVC plastic pipe, concrete or ceramic pipe can be used, it is necessary to ensure the size, not broken, clean surface for better connection with the equipment.

3. The following recommendations should be taken into account when preparing the soil:

a) Clearing of land for construction: levelling of the land, drainage, clearing of plants on the land.

b) Plan the right place to concentrate soil, construction material, calculate the amount of soil to bury later.

c) The design and soil analysis at the installation site, determine the size of the pit and the way of digging. The decomposition pond is constructed first, the other components are installed later.

d) The location of the components on the ground is determined with the digester center, from this center, the circumference of the pond is calculated on the plane to excavate. If the soil is weak, this circumference must be opened.

4. Measurements for construction.

With small plants, manual digging is applied, with larger equipment, manual and machine digging can be combined. If the soil is solid and dense as clay, solid and the pit depth is less than 3m, the ground water level is lower than the bottom of the pit, the pit wall can be vertical with the pit diameter of 20-30cm larger than the diameter of the pond. If there is a water source crossing the pit, it is covered with clay. If the depth is more than 3m, open the hole mouth with the corresponding slope. The excavation should be as shown in the figures.

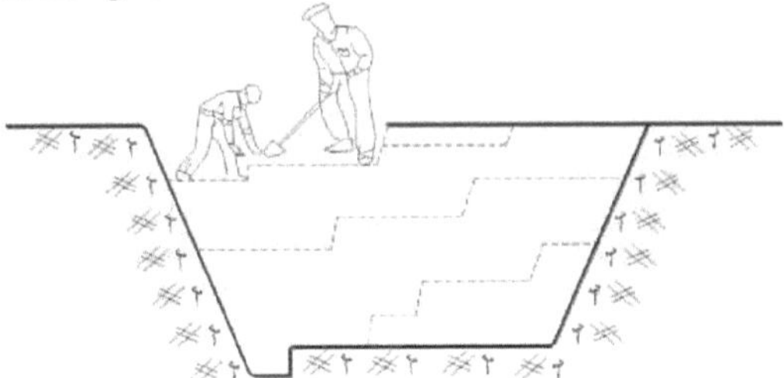

Fig.3.11.-Manual excavation measure.

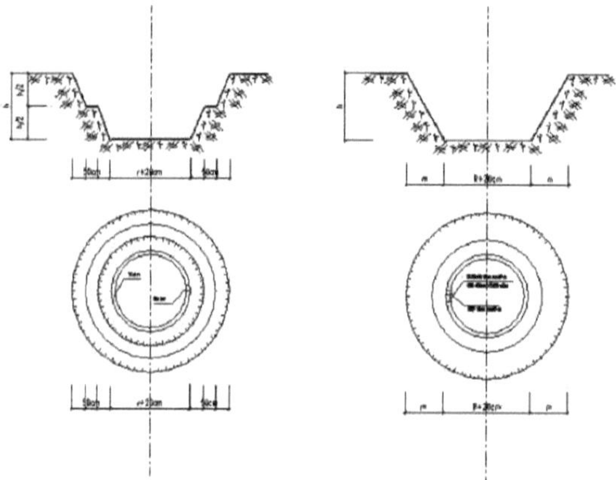

Fig.3.12.- Hole pattern for the base

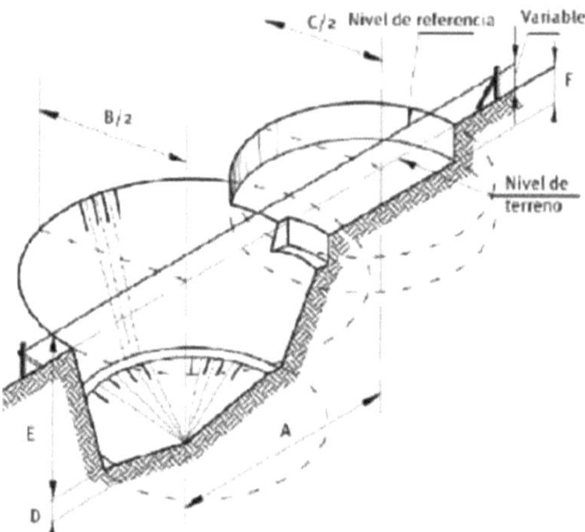

Fig.3.13.- Excavation for typical digesters

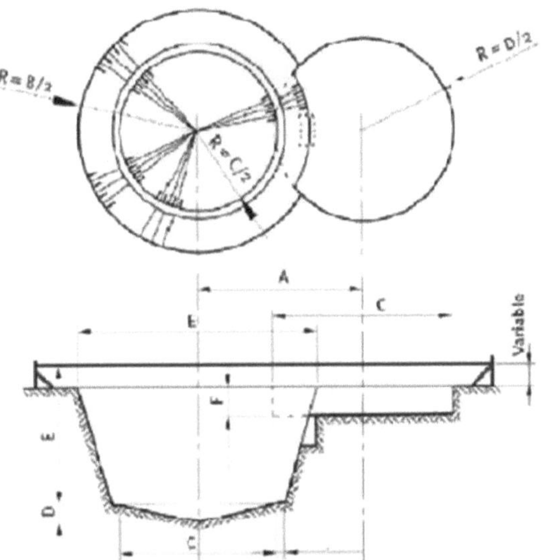

Fig.3.14.- Excavation for typical digesters

With other types of soil (soft elastic clay, clay mixed with sand...) it is necessary to open the opening of the pit with the corresponding slope or apply the measure of staking to avoid the collapse of the pit wall. In the above-mentioned cases, if there is a high water source, the bottom can be raised or a hole can be made to accumulate water, drain and keep the bottom always dry at the time of construction. The geological condition of the construction site should be well analyzed before applying the appropriate forms of structure and construction.

5. Base construction

The base can be constructed of brick or concrete according to geological conditions and pond volume. The base construction with brick can be executed with the steps:

a) The bricks are laid in a circle with the same center, do not repeat the channel (space between bricks) and ensure that between bricks there is mortar, if the channel is horizontal, the space is 8-12mm, if the channel is vertical, the space is 10mm.

b) With wet soils or ground water, put 1 to 2 layers of nylon underneath, build on them and keep the base dry for at least 24 hours after building.

c) After construction, avoid steps over it or put boards over it to pass when the mortar is not solid.

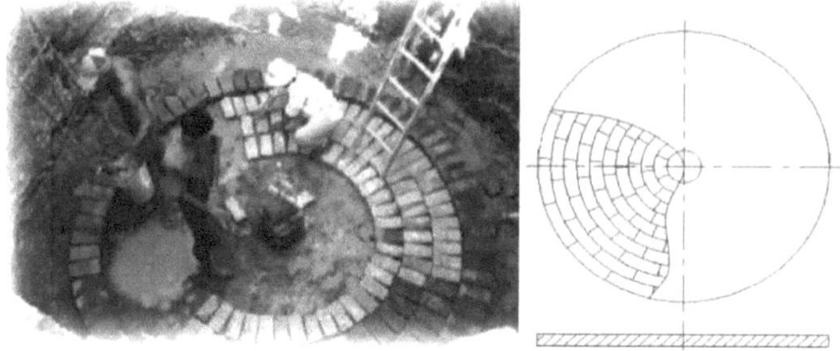

Fig.3.15.-Bottom of digester built with brick.

Fig. 3.16.- Central axis and wooden bracket (conical slab foundation)

With the weak soil, the concrete base should be built with pieces of bricks or stones, gravels of density 10 to 20cm. The concrete should be well mixed and ensure the component of 1 cement / 3 yellow sand / 6 gravels (the pressure to withstand is 150 -200kg/cm^2). The concrete should be compressed to increase the solidity, avoid voids and then cover the base with cement plus sand at a ratio of 1:3. The concrete must be kept dry for 24 hours after the operation for it to become solid. When the base is solid, the construction of the pond begins.

Fig. 3.17.-Bottom of the digester constructed.

Fig. 3.18.- Central shaft and wooden bracket (beginning of the construction of the biodigester body cylindrical wall)

The base must be rigid in bending in order to transmit the edge loads to the total surface.

6. Construction of the pond wall.

Brick wall size 110: the wall has the density equal to the brick width, the bricks from top to bottom is at an alternative of % of brick length.

When laying the brick, pour mortar on a point of brick and put it on the line with it a little bit down so that the mortar fills the vertical channel (space between bricks in a line). After pouring mortar on the constructed line, lay bricks in a line with the other bricks already laid. The construction of spherical roof requires high technical skills, it is necessary to follow the following methods of orientation:

a) Center and diameter of sphere (to be removed after construction): build a small brick pillar to hold a pole in the center of the pond bottom (drawing); determine the center of the sphere on this pole. It is best to mark this center with a nail on the pole. The distance from the center marked on the pole to the bottom of the pond is equal to

half the pond diameter.

b) Use a rope to determine the radius: tie a point of rope to nail (must tie so that the rope can rotate). Determine the pond radius and mark it on the rope. In the construction of the wall, always use this string with its marking to ensure that the position of each brick is at a certain distance from the center already defined, as shown in the figure:

Fig.3.19.- Construction of the pond wall

In the construction of the arched wall, starting from a solid base, it is built in each line from the bottom to the top as a closed circle. When laying each brick, use the above-mentioned string to check the distance and the angle of declination of this brick to ensure the correct arc shape. Before building the first line, use the radius determination string to mark a circle at the base and build the first line to this circle. The first lines can be collapsed if the mortar does not dry, after the first line, use bars or brick to support outside, avoid collapse. The first lines should be well plastered because it is the least resistant position with water ingress and egress. The plaster layer must be dense and compressed. It is plastered in the shape of an arch, without angle, a bottle can be used to plaster better in the shape of an arch.

Use a rope or a cramp tied to the brick that needs to be supported, the other end of the rope or cramp is tied to a fixed place or a heavy object to counterbalance the newly placed brick in each line until the mortar sets solid (see figure). The first brick of each line should be supported until the entire line is finished.

The construction of the dome in the shape of an arch: it is an easy task that requires a skilled builder and the measurements can be applied:

a) Choose bricks with a good shape, soak them in water.

b) Use mortar with a high adhesive character: mortar with cement, lime and

sand at a ratio of 1:1:4. Or cement with sand at a 1:3 ratio.

Fig. 3.20.- Gwa de cuarton for the construction of the cupola.

The construction of the dome should be divided into several stages. On the first day you can build the first two lines after the construction of the base. Wait a day for the base to dry, continue construction until the moment when it is necessary to use a support tool. In that part, it is obligatory to build every two or three lines and wait until it becomes solid. During the waiting time, it is possible to build regulation tank or inlet material dissolution tank.

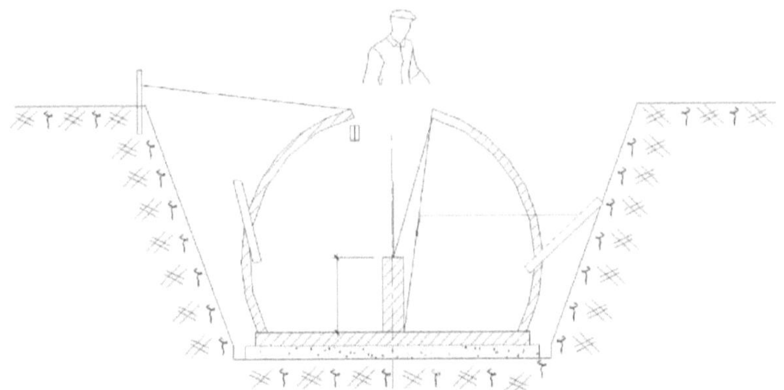

Fig.3.21.-Construction of the pond wall to a spherical shape.

Fig. 3.22.- Iron hooks to aid the bonding of bricks.

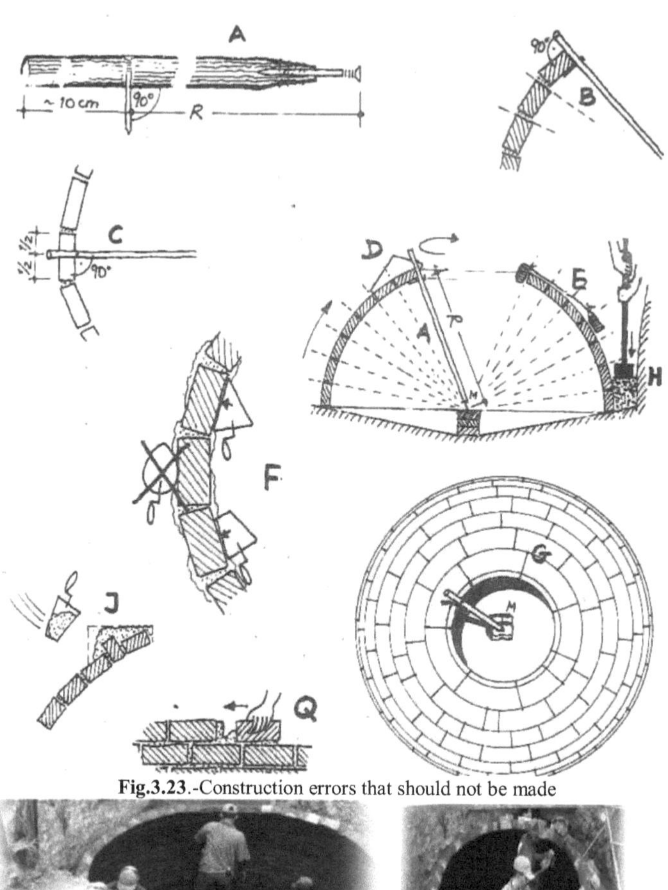

Fig.3.23.-Construction errors that should not be made

Fig.3.24.-Construction details

7 **Inlet** pipes

When the pond wall reaches the position of the pipes, put the pipes to the designed position, these positions are very easy to get water out later, it has to be executed very carefully to avoid the complex repair originated in the later. The point of connection of the pipe with the pond must be consolidated with mortar, this mortar must not be too wet and it is necessary to plaster again when it dries, especially the part below and the surface behind the pipe.

The loading pipe should be straight. The axis of the tube should point as far as possible to the center of the digester. The inlet of the loading pipe should be at the top, so that the settling sand does not clog the loading pipe. Sand and stones should settle in the mixing tank.

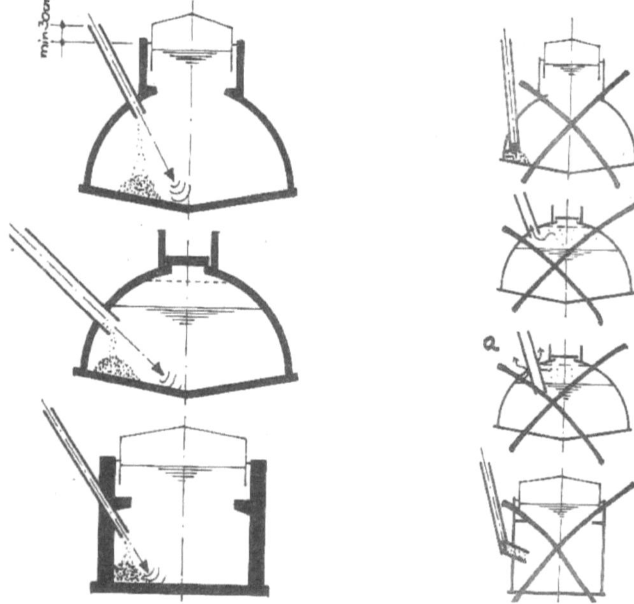

Fig.3.26.- Correct form

The mouth of the loading pipe should be 3 - 5 cm above the tank floor. Cylindrical construction is the best and most economical shape.

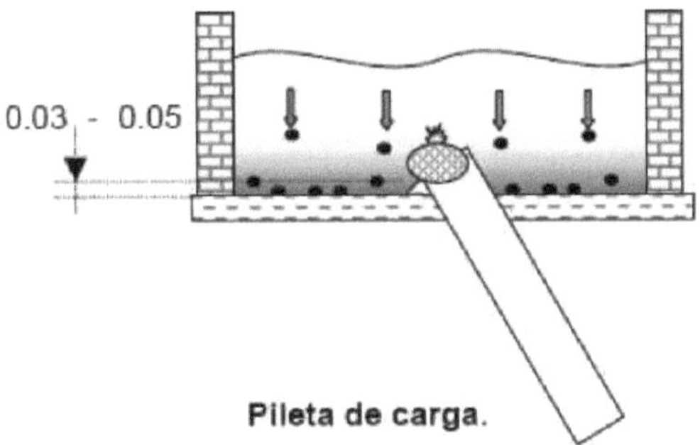

Fig. 3.27. - Loading basin

If the mixing tank is filled and covered with a metal plate until the evening, the fermentation sludge is warmed by the sun, more than if it were uncovered. Then in the evening the digester is loaded.

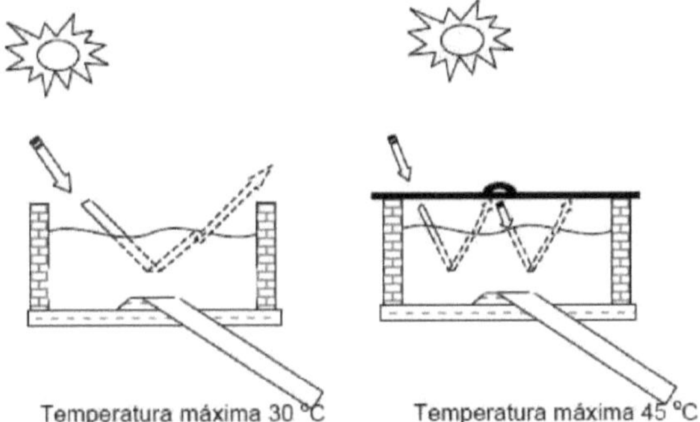

Fig.3.28.- Influence of sunlight

8 Discharge pipe

The mouth of the discharge tube should be as low as possible. Otherwise a lot of fresh, unfermented fermentation material escapes. The height of the upper mouth of the discharge pipe determines the height of the fermentation sludge level. This mouth should be 3 cm below the upper edge of the wall.

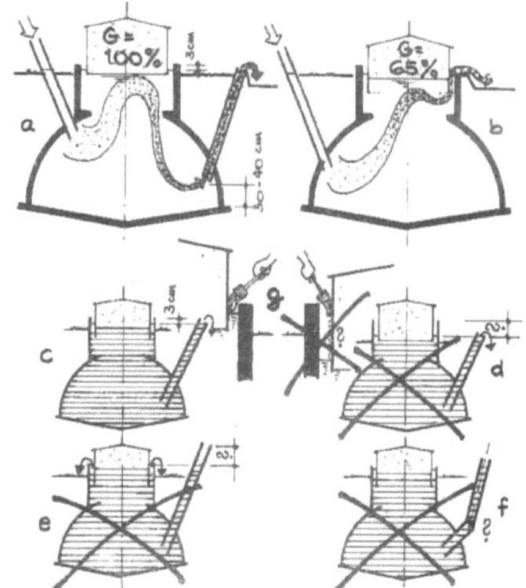

Fig.3.29.- Correct Installation

9 Construction of the neck of the digester.

The so-called "neck" of the biodigester is built on top of the beam. The neck is a cylindrical wall composed of two sections and is built with bricks in brick in the lower part on the beam up to the height of the cornice on which the lid will rest and in rope in the upper part. One course before the end of the horizontal part is placed the conduction pipe of the biogas oriented towards the place of consumption. The mortar used is similar to the one used in the construction of the cylindrical wall of the central body of the biodigester.

It is done in two steps:

    a) First: after building the dome wall, start the construction of the neck base (lid support).

    b) Second: The rest of the collar is executed. The plastering work has to be finished in the pond before in order to have enough light and a comfortable entrance and exit of the pond.

Fig. 3.30.- Neck of biodigester

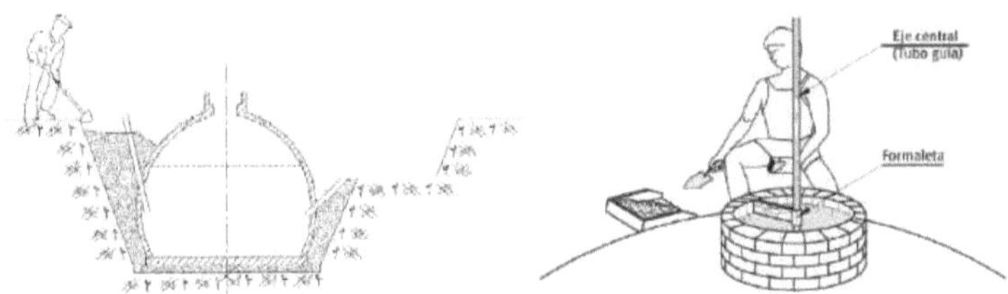

Fig.3.31.-Digester neck construction.

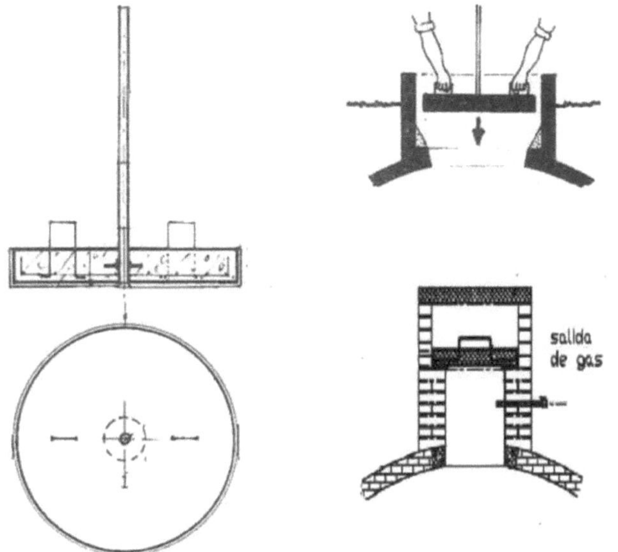

Fig.3.32.- Details on the inlet hatch or lid

Fig. 3.33.- Detail of neck closure

10 Construction of the regulation or compensation tank.

The construction of the regulating or equalization tank is carried out in the same way as the decomposition tank, the basic thing is to set the outlet level as it is shown in the design. It is possible to start the construction of the regulating tank after the installation of the outlet pipe and to reach the part of the decomposition tank dome (as mentioned above), or it is possible to finish the digester before starting the regulating tank.

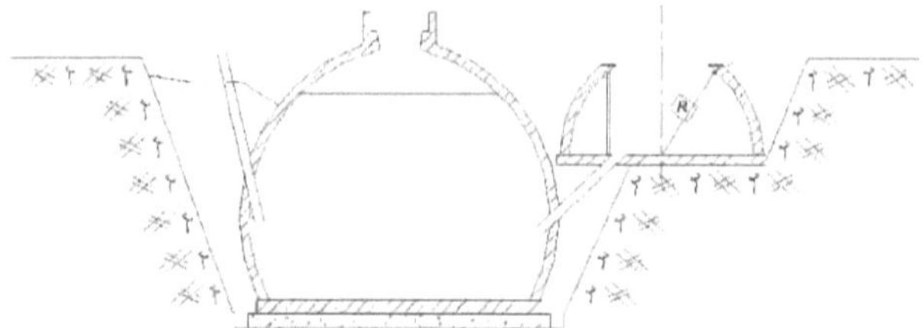

Fig.3.34.- Supporting pole to build regulation tank

11 Repel and protect against water absorption.

This work is very important to prevent water absorption, especially in the inner layer of the tank. To save materials, it is not necessary to plaster the outside of the tank. The mortar to be used should be fine sand, well mixed, in proportions of 1 cement/3 sand or 5 cement/1 lime/15 sand.

The general requirement is to keep the density stable, to compress the layer well and to avoid the formation of angles on the surface. Only a layer of 1 cm is necessary, compressed and painted with cement dissolved in water. In the inner part of the decomposition tank, the following steps must be followed:

1. Clean the surface for repainting if it is dirty.
2. Use pure liquid cement or a mixture of 5 cement/1 lime to cover the surface to be plastered.
3. Plaster a 0.5 cm layer of mortar (1 cement/3 sand). Wait for it to dry and compress it with a trowel. Wait for one to two hours for the mortar to dry and then apply another layer as explained above, continue with the third layer and cover with pure liquid cement.
4. Paint with anti-absorption Kquido: use cement Kquido mixed with anti-absorption element to paint over the repainted layer, the operation is repeated 3 times, as long as the previous layer is dry. The main part of the tank dome and the part of the pipes up to the neck of the tank are painted with this liquid.
5. The cover must be concreted from the beginning to ensure the necessary solidity. The concrete is mixed in proportions of 1 cement / 2 sand / 3 gravel, the lower and surrounding parts have to be covered with mortar and smoothed. Then the 3 layers of anti-absorption paint are applied as in the interior of the decomposition tank.

12 Pressure test

In biodigesters, gas storage is done inside the biodigester. Inside, pressures ranging from 1.0 to 1.45 meters of water column (0.1 to 0.145 kg/cm^2) are present, depending on the capacity of the biodigester. The impermeability of the system is verified by means of the pressure test, which is performed after completion of the biodigester construction, to allow the masonry to set and ensure adequate resistance.

The pressure test consists of simulating the real Ionic operation of the plant and is carried out over a period of three days. For this purpose, a compressor or an internal combustion engine (for example, that of a car) must be available. The air or exhaust gases from the velncle are introduced into the biodigester. The procedure for the pressure test is as follows:

On the first day, the biodigester must be sealed and simultaneously with this, the filling of the plant with water is started through the slab load register that reaches the height of the lintel of the window. The concrete cover is placed on the neck with the following recommendations:

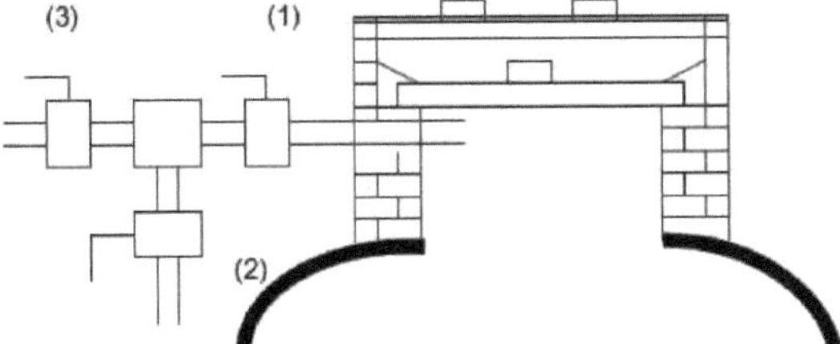

Fig. 3.35.- Piping installation for pressure test.

a) The surface of the neck and lid should be cleaned and soaked with water.
b) For the seal, clay or clayey soil is used, sifted and mixed with cement: in a cement-clay ratio 1:8 and little water to obtain a moldable mass that is placed on the surface of the cornice of the collar; the cover is installed making sure that the height between it and the collar is equal around the entire perimeter. Then a person must stand on it, exerting a uniform pressure to compact and achieve a good adhesion of the clayey mass. Then the space between the neck and the lid is filled in a compact form and finally to reinforce the neck a poor mixture of cement-sand in proportion 1:8 is placed around the lid. Two hours later the neck is filled with water.
c) Installation of piping and fittings for the test: the piping and fittings are installed as shown in Figure 4.

   (2) for the pressure gauge and another one for the air inlet (3) or engine exhaust gases. To operate the system properly, the valves indicated in (1) (2) (3) must be installed.

On the second day, the biodigester should be filled with water until it overflows the compensation lagoon. The pressure gauge is connected to the valve (2) and the valve (1) is also left open. The valve (3) must be closed to prevent the air in the dome from being displaced. When the pressure increases due to air compression by the water, it is advisable to check the installation with soap and water to detect leaks and to observe if there are water bubbles in the neck.

The pressure indicated by the pressure gauge should be measured and recorded when the water level overflows into the compensation pond. The pressure behavior should be observed for one hour and the valve (1) should be closed. It is possible that the pressure decreases in this first phase, because the construction is still absorbing water.

To check the plant pressure, first open valves (2) and (3) to depressurize the pressure gauge. With valve (3) closed, slowly open valve (1), measure and record the pressure registered on the pressure gauge.

On the third day, air is introduced into the plant. If the test is carried out with an internal combustion engine, it should be noted that the high temperature of the gases may melt the PVC pipe. It is therefore recommended to use a piece of metal pipe at the connection with the exhaust of the vehicle and to cool it with water. Install the compressor or internal combustion engine to introduce gases into the plant, making the connection with the valve (3). The valves (1) and (2) are opened and the engine is turned on so that gas enters until it reaches maximum pressure, closes the valve (3) and turns off the compressor or engine. The pressure is observed at this time and checked the following day. The test is considered satisfactory when the pressure is maintained for a period of 24 hours. If pressure losses are greater than 10%, you should check the biodigester and seal the leaks.

13 Selection and installation of gas pipe and fittings

The gas produced in the biodigester must be piped to the sites of use. PVC pressure piping is suitable for this purpose, although certain precautions must be taken.

PVC pipe has the following advantages:

    a) It is corrosion resistant.
    b) Its walls are smooth.
    c) It is lightweight and easy to install.
    d) It is economical

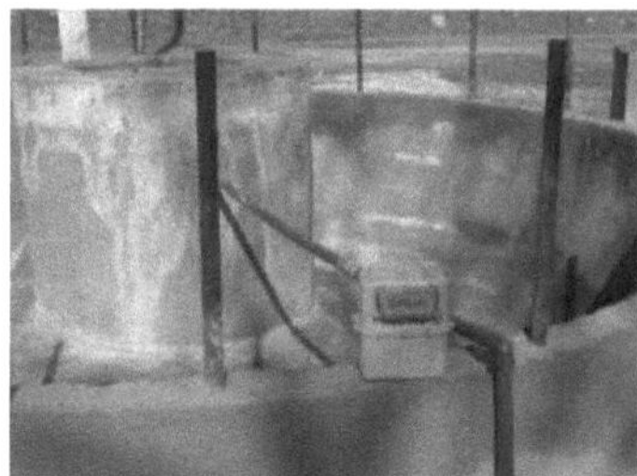

Fig. 3.36.- Piping installation.

As a precaution, it should be installed underground to protect it from the sun's rays and sufficiently buried to withstand the transit of heavy vehicles and the passage of livestock. The diameter of the gas pipe required in the installation depends on the distance from the plant to the place of use of the biogas, the maximum flow of biogas required and the admissible pressure loss.

Pressure losses of up to 50 mm of water column can be tolerated in plants with a fixed dome and up to 10 mm of water column in plants with a floating hood or independent gasometer. The maximum gas flow is obtained by adding the consumptions of the appliances operating simultaneously and causing the maximum gas consumption at a given time. Table 3.3 allows to select the pipe diameter:

Table 3.3.- Pressure losses in biogas pipelines (PVC)

	LOSSES FOR EVERY 10 METERS OF PIPE (meters water column; mm/c.w.)				
Flow rate (m³ /h)	¾" (17 mm)	⅞ " (23 mm)	1" (34 mm)	1.5" (43 mm)	2" (68 mm)
0,5	1,0	0,3	0,1		
1,0	2,5	0,7	0,2		
1,5	4,6	1,2	0,4		
2,0	7,0	1,8	0,6		
2,5	9,9	2,5	0,8		
3,0	13,1	3,3	1,0	0,2	
4,0	20,7	5,2	1,6	0,3	
5,0	29,6	7,4	2,0	0,4	
6,0	39,7	9,8	2,9	0,6	
7,0		12,6	3,7	0,7	
8,0		15,7	4,6	0,9	0,3
9,0		19,0	5,6	1,0	0,4
10,0		22,6	6,0	1,3	0,5

14 Water traps and pipe slopes.

The gas from a biodigester usually comes out with water vapor. Some of the water vapor can condense in the gas piping causing blockages if the piping has not been properly installed. Condensed water in the piping should drain to low points where water traps are located.

The number of water traps required varies according to the topography of the terrain and the length of the route. A slope of 2 % is sufficient for the installation of water traps. There are different types of water traps, of closed type with a ball valve to purge the condensed water after a certain time of operation (this type is the most used in the biodigesters installed in the province of Matanzas) and of open type like the U-trap; in these traps the height (h) must compensate the pressure of the plant and also about 30 cm of overpressure.

It is recommended that a ball valve be installed near the biodigester to shut off the gas flow in the event of a pipeline or appliance repair. A pressure gauge such as the one described in the pressure test is useful to install near the places where the gas is used, as it provides an indication of the amount of gas in storage. It is desirable to have a ball valve upstream of each appliance, however, this is usually uneconomical; at least one valve should be placed upstream of the stove for safety.

15 Start-up and operation of biodigesters.

Start-up: This is the most important moment to achieve a good operation of the biodigester, if this is not correct it will not be possible to achieve an adequate operation and has caused in many cases the loss of interest of the beneficiaries in the use of these systems. The start-up of the biodigester is carried out with the available load material; pig and/or cattle manure (which are the most frequent and of greater availability in the rural areas of our country), which must be inoculated or introduced into the system with an adequate manure-water mixture to achieve a rapid stabilization of the digestion process and biogas production.

If the pressure test was satisfactory, start the system by operating fresh manure-water mixtures in preferably equal proportions or by using water up to a maximum of three times the amount of manure (1:3). In places where it is not feasible to rationalize water, it is recommended to collect fresh manure manually before washing and deposit it in the load register or near the washing channel and dilute it with the necessary water. This management allows to reduce water consumption.

Operation: In order to have a growth of methanogenic bacteria and continuous production of biogas, the biodigester must be loaded daily. To prevent inert material such as sand, stones, and others from entering the biodigester, it is necessary that the bottom of the load register has a negative slope towards the filling direction of the biodigester, or otherwise build a sand trap before the biodigester, which should be cleaned periodically. During the loading operation, check that the pipe that conducts the manure-water mixture to the biodigester is open. Once the loading operation is finished, plug the feed pipe to prevent different materials or rainwater from entering the system. Verify that the biogas conduction pipe does not contain water as it impedes the passage of biogas and its utilization. If this occurs, check the water trap. Weeds should not be allowed to grow around the plant.

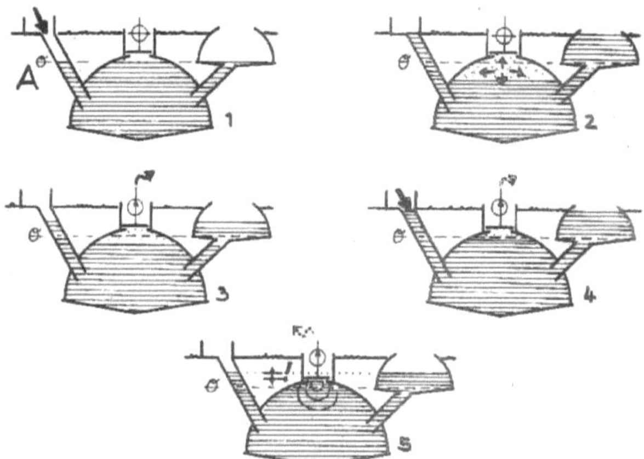

The plant is loaded; 2.-The gas produced pushes the fermentation sludge into the compensation tank; 3.-Before it spills, gas is extracted; 4.-Even without having extracted the sludge, it is loaded again; 5.-If gas continues to be consumed, the fermentation sludge is pushed back and the plant stops working.

Fig.3.37.- Operation forms

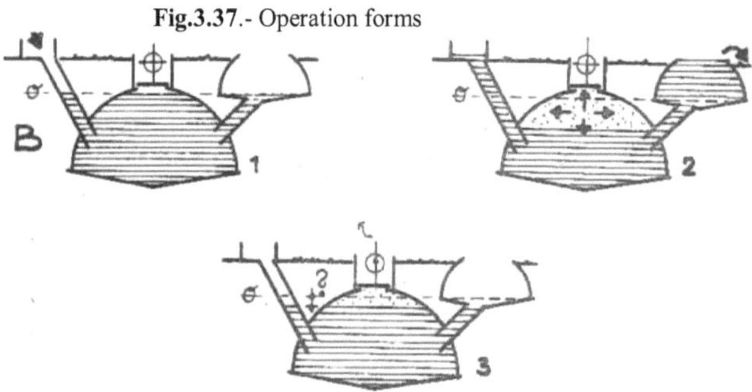

1 The plant is loaded to the zero line; 2.-The produced gas pushes the fermentation sludge; 3. The fermentation sludge has flowed down the overflow. The gas below the steel line is also unusable.

Fig.3.38.- Operation forms

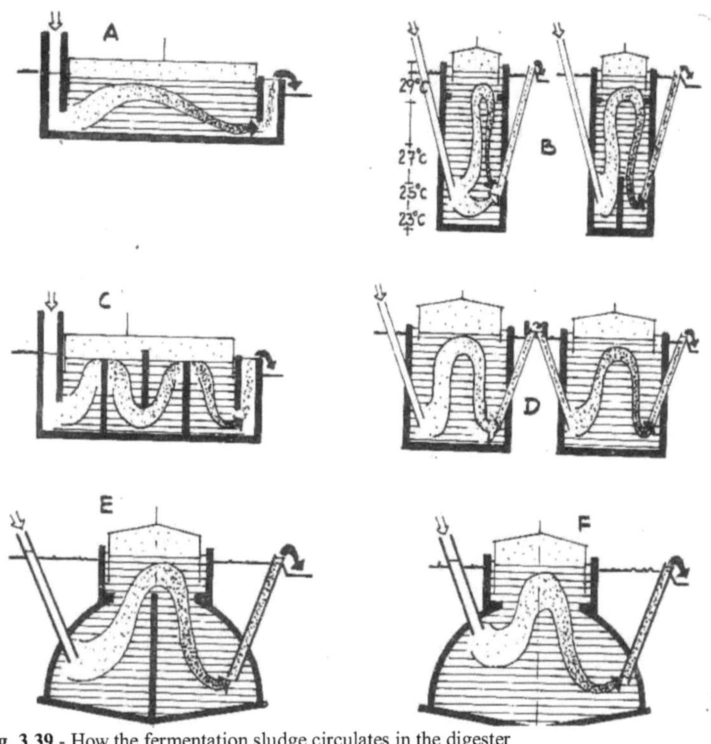

Fig. 3.39.- How the fermentation sludge circulates in the digester

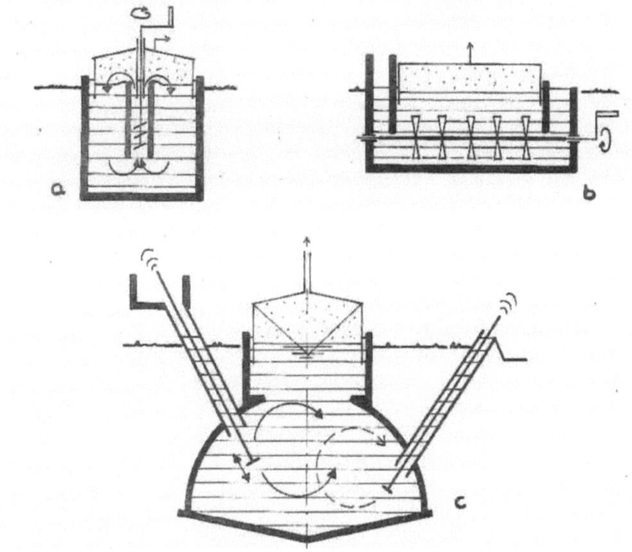

Fig.3.40.- Stirrer Operation

16.- Maintenance

A biodigester requires general maintenance when there are problems in its operation. These problems are manifested when:

a) The entry of the feed material is hindered due to the accumulation of solids in the feed pipe or inside the biodigester.

This causes low biogas production; to solve this situation you should remove the clay seal from the lid; remove the lid and let the biogas come out, DO NOT SMOKE; if you have a motor pump, evacuate the solids from inside the biodigester or use buckets. Before entering the biodigester, make sure that it is well ventilated: otherwise asphyxiation may occur, remember that there is no oxygen inside, which is why we insist on the need for proper ventilation before a person enters the biodigester. Completely discharge solids and sediments from the bottom and clean the compensation lagoon. If possible, store a part of the sludge, which can be used again to restart the biodigester.

b) There is biogas leakage through the dome, the neck or the junction zones between these two parts or between the biodigester cylindrical body and the dome.

This situation is determined by a loss of biogas pressure, which can be seen in a deficient functioning of the equipment that uses it, such as stoves, internal combustion engines, heaters, etc. To determine the presence of the leak, a mixture of water with soap or detergent is prepared and applied to the surface of the dome, the neck cylinder and the lid. The area where the biogas is leaking is determined by the presence of bubbles. Then the biodigester must be unloaded with a procedure similar to the one described above and proceed to the internal cleaning of the area where the biogas leakage was determined. The stucco is lifted in that area, it is chopped with the pickaxe and the stucco is put back on, moistening well the area where it is going to be applied and the surrounding area. To achieve a good adhesion both to the affected area and the surrounding area, before applying the stucco, sprinkle pure cement on the area, which contributes to the union between the two bodies, i.e. between the surface to be stuccoed and the stucco. This sealing can also be done by using other additives such as those indicated in the explanation of the ways to achieve the waterproofing of the dome and the neck, always bearing in mind that they are substances that do not leave any residue of chemical character, because this can affect the bacteria and cause the biodigester not to work.

c) The biodigester does not produce biogas despite being fed on a regular basis.

This problem can be caused by the presence inside the biodigester of substances that inhibit or impede the development process of the bacteria inside the system. This phenomenon is known as "biodigester poisoning" and is generally due to carelessness that causes the entry of harmful substances such as hydrocarbons (petroleum derivatives), antibiotics used in the treatment of animals or people, as well as other substances that can affect certain parameters such as the pH of the medium, as well as other substances that can affect certain parameters such as the pH of the medium, including lime, residues of inorganic acids such as sulfuric acid (acid from accumulators), chlorludric acid (salfuman) and others. Care should also be taken to ensure that no inorganic fertilizer residues enter the biodigester that may increase the presence of metal cations whose concentrations are detrimental to the microorganisms and inhibit the anaerobic digestion process.

d) Low biogas production and fetidity of the final sludge.

Generally this problem is due to deficiencies in the biodigester feed system, mainly overfeeding, i.e., the amount of manure fed to the system is greater than the carrying capacity of the system. This occurs when the biodigester is not properly designed and is smaller than what is needed to assimilate the amount of waste generated. When this phenomenon manifests itself, the biodigester should be stopped feeding for several days and the contents of the biodigester should be stirred with an agitator to avoid the formation of crusts. If after several days the problem still persists, the entire contents of the biodigester should be evacuated as explained above, cleaned well and reloaded. It should also be checked that there are no cracks in the constructive structure, because the entry of air into the system causes the destruction of the anaerobic system and the process occurs by aerobic decomposition, which produces stench (unpleasant odors).

e) The biodigester remains idle for a relatively long time.

This can occur when, for various reasons, the system lacks feed, i.e., the required amount of waste is not available to keep it functioning properly, which may be due to several factors, among which are: that the animal breeding system (pigs, goats, sheep, etc.) presents problems due to diseases that affect the animal mass, economic problems of the producer, breaks in the consumption system, or others. In these cases the interior content of the biodigester, after having been transformed by the action of the methanogenic bacteria, is transformed into an inadequate material for the feeding of these, which causes them to die. This material tends to harden and it is necessary to proceed to the total evacuation of the content and to clean the system well. To carry out this action, it is recommended to add a certain proportion of water to soften the sludge and remove it by the means recommended in the first aspect addressed in this chapter referring to maintenance and taking into account the recommendations given in this regard.

The frequency of maintenance work depends on the specific conditions of each digester:
    a) Removal of cream and supernatant.
    b) Solids and sludge removal in the surge tank.
    c) Checking the condition of the pipes.
    d) Drainage of water traps and low points of the pipelines.
    e) The biogas plant must be a pleasant and welcoming place, so painting and maintaining cleanliness is as necessary as its satisfactory operation.

Bibliography.

1. De Le m u s C h e r n i c h a r o , C a r l o s A u g u s t o . A n a e r o b i o s Reactors . Volume 5 . Federal University of Minas de Gerais.2009. Belo Horizonte. Belo Horizonte. Brazil.
2. Guardado Chacon, Jose Antonio. Biogas technology. User's manual. Editorial CUBASOLAR. 2006. Havana. Cuba.
3. Guardado Chacon, Jose Antonio. Design and construction of simple biogas plants. Editorial CUBASOLAR. 2007. Havana. Cuba.
4. House, David William. The Complete Biogas Hanbook. 2009
5. Lugones Lopez, Barbaro. Analysis of biodigesters. 2001. Cuba.
6. Recio Recio, Angel Amado and collaborators. Energetic use of the biogas from wastes of the Brewery Hatuey in Santiago de Cuba. ISBN 978-959-16-1246-5. XV International Scientific Congress of the National Center for Scientific Research. 2010.
7. Recio Recio, Angel Amado and collaborators. Studies and projects carried out on the energetic use of waste using the technology of anaerobic digestion of the center of studies of energetic eiiciencia. Paper presented at XI DAAL.XI Latin American Workshop and Symposium on Anaerobic Digestion. November 24-27, 2014. Havana. Cuba. ISBN-978-959-261-470-3.
8. Sanchez Rodriguez, Jose V. Introduction to biogas production. 2005. Cuba.

a. Sasse, Ludwig. The Biogas plant. Sketch and details of simple plants. ISBN 3-528-02010-5. 1984. Federal Republic of Germany.
b. Silva Vinasco, Juan Pablo. Biogas Technology. University of Valle. Faculty of Engineering. Colombia.

ANNEXES.

Annex 1

Number of pigs, digester volume, biogas production per day and use of the **digester.**

Number of Pigs	Digester volume, m³	Biogas production, m /day³	Uses of biogas	
			For cooking	Generate electricity, kwh
10	4	1,4	One burner for 5 hours	2,24
20	4	2,9	One burner for 10 hours	4,64
30	8	4,5	Two burners for 7 hours.	7,2
40	12	5,9	Two burners for 9hours.	9,44
50	12	7,4	Two burners for 11 hours	11,84
70	22	10,4	Four burners 8 hours	16,64
100	22	14,8	Four burners 11 hours	23,68
200	48	29,5	Two industrial burners 8 hours	47,2
300	70	44,5	Three industrial burners 8 hours.	70,72

The following requirements were considered for the calculations:

1-Fattening pigs with an average weight of 55 kg.

2-Water for cleaning the corrals should not exceed 8 liters per animal.

3   The hydraulic retention time was considered to be approximately 20 days.

4   . -The construction of the digesters must be carried out by masters with experience in this type of work.

Annex 2

Fixed dome digester materials.

NO.	Materials	U/M	Capacity of the digesters m³							
			4	8	12	15	22	35	48	70
1	1/4" sprocket	m	22	27	30	35	40	108	115	136
2	3/8" sprocket	m	58	64	70	75	80	156	164	250
3	1/2" sprocket	m	81	92	97	110	120	138	147	168
4	5/8" sprocket	m	1.5	1.5	1.5	1.5	1.5	1.5	1.5	1.5
5	Ladrllo of clay of 25x12x7 cm.	u	368	511	670	772	900	1216	1420	1875
6	Concrete block 15x40x20cm	u	188	260	320	387	435	600	800	900
7	Portland Cement 250 (50Kg)	Saco	30	40	50	60	68	100	125	135
8	Portland Cement 350 (50Kg)	Saco	2	2	3	3	4	5	6	8
9	Washed Sand	m³	3.25	3.5	4	4.7	6	8	9.5	11
10	Silica Sand	m³	0.5	0.5	0.5	0.5	0.5	0.5	0.5	0.5
11	3/4" stone	m³	2.3	2.5	3	3.3	4	5.6	6.3	7.5
12	Lime slaked	Saco	0.5	0.5	0.5	0.5	0.5	1	1	1.5
13	6" asbestos cement pipe	m	1.5	1.7	1.8	1.8	3	3	3	3
14	Cardboard board 30 cm. wide	m	18	23	25	30	38	42	50	60
15	3/4" ball valve	u	1	1						
16	3/4" polypropylene piping	m	1	1						
17	1" ball valve	u			1	1	1			
18	Polypropylene pipe 1".	u			1	1	1			
19	1 1/2" ball valve	u						1	1	1
20	1 1/2" polypropylene pipe	m						1	1	1
21	Cable tie wire	Kg.	1.5	2	2.5	3	3.5	3.8	4	4
22	3" nails	Kg.	1.5	2	2.5	3	3.2	3.3	3.5	3.5

Annex 3.
Biogas digester construction pianos.

4.9 m BIOGAS DIGESTER[3]

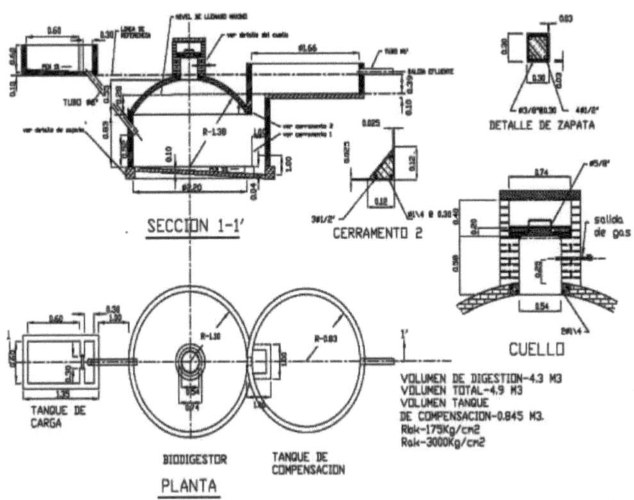

DIGESTOR DE BIOGAS DE 8 m³

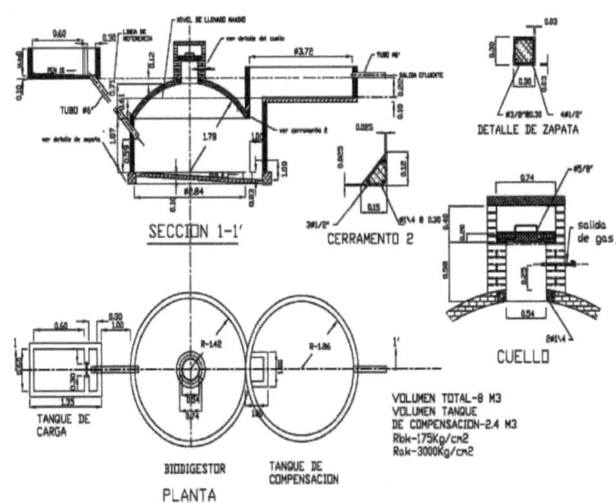

DIGESTOR DE BIOGAS 12 m³

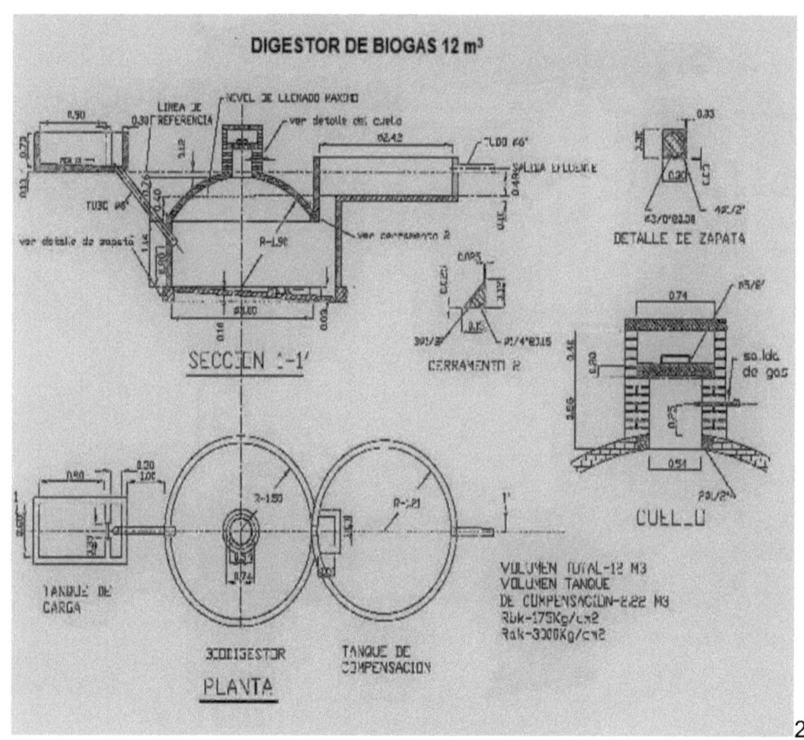

m BIOGAS DIGESTER³

DIGESTOR DE BIOGAS DE 48 m³

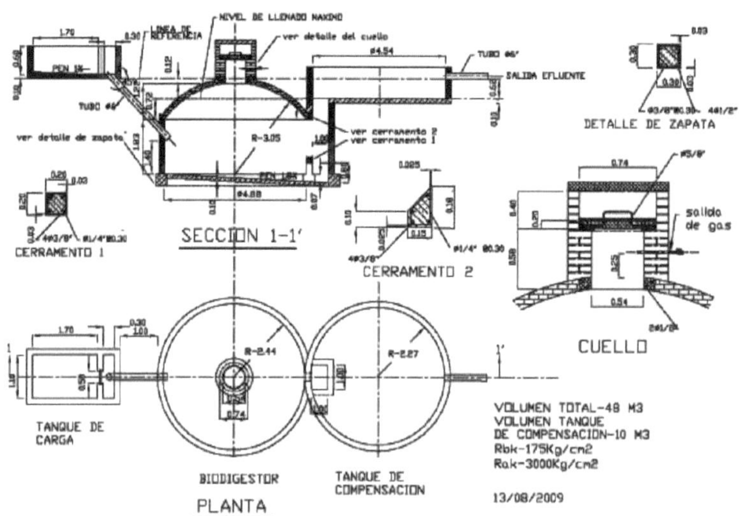

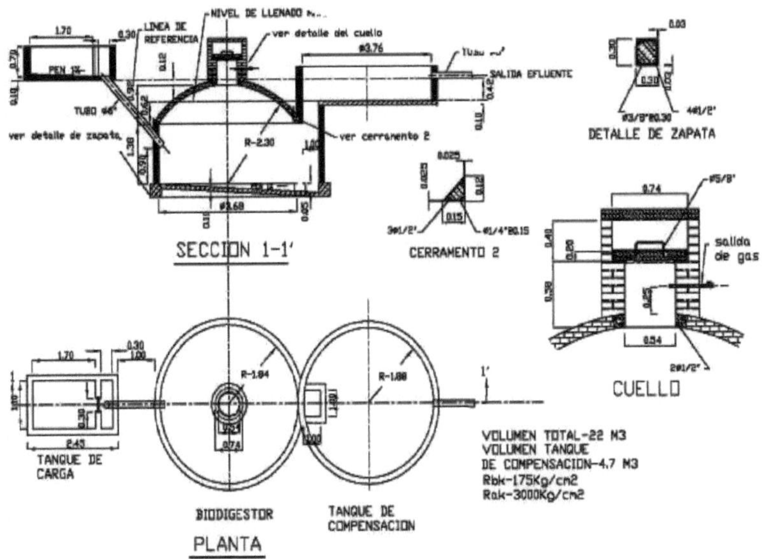

DIGESTOR DE BIOGAS DE 70 m³

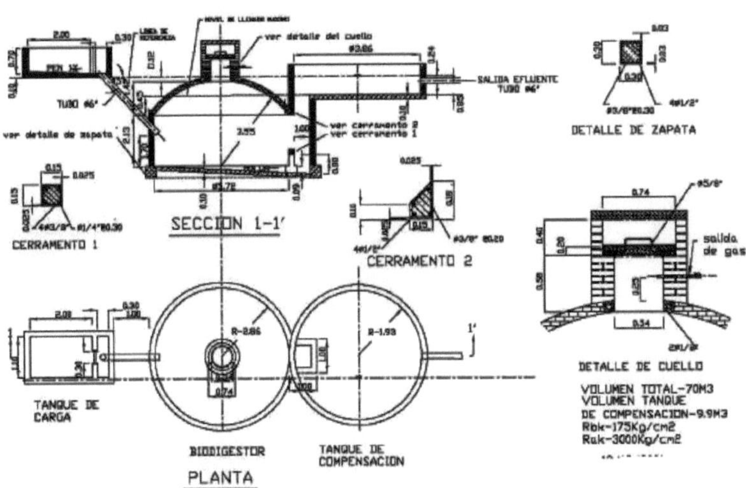

18,45 m BIOGAS DIGESTER³

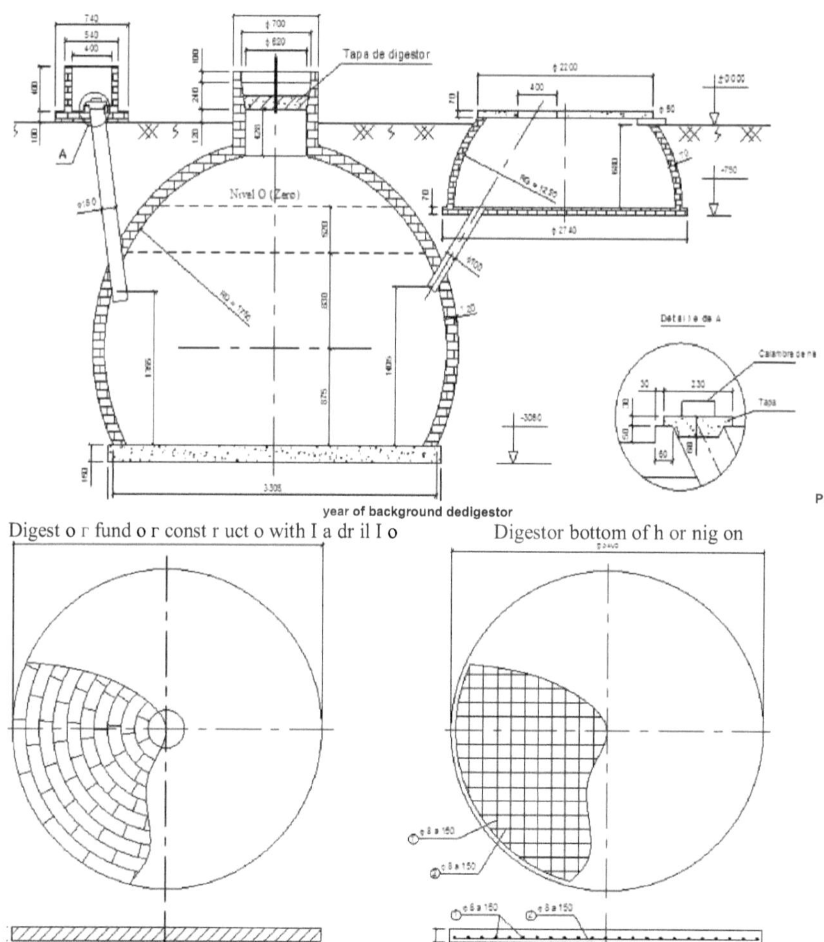

Digest o r fund o r const r uct o with I a dr il I o

Digestor bottom of h or nig on

Cuel l o de dig est or

Fondo de t anque de r egul acio n de l adr il l o

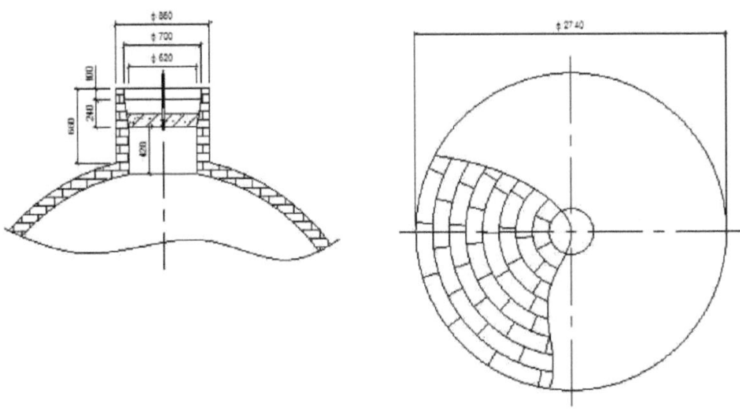

Dis eno de t apa de t anque de r egul acio n y d lg est or

Est r uct ur a de hier r o de t apa de t anque de r egul acio n
(Dividido 2 dos partes)

Est r uct ur a de hier r o en l a t apa de dig est or
(una tapa)

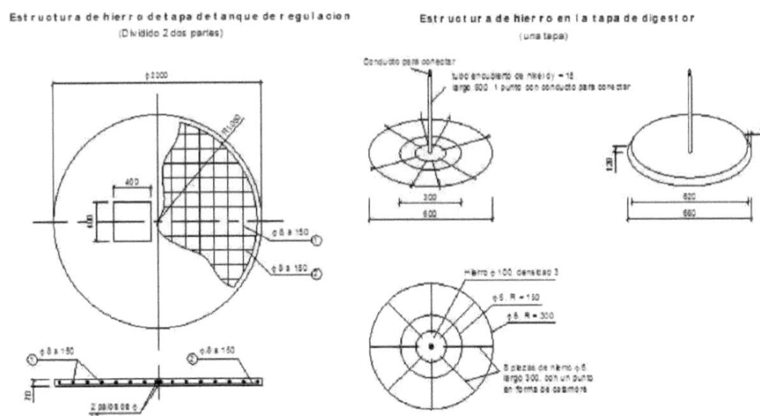

Profile of inlet-digester-digestion tank materials
±360

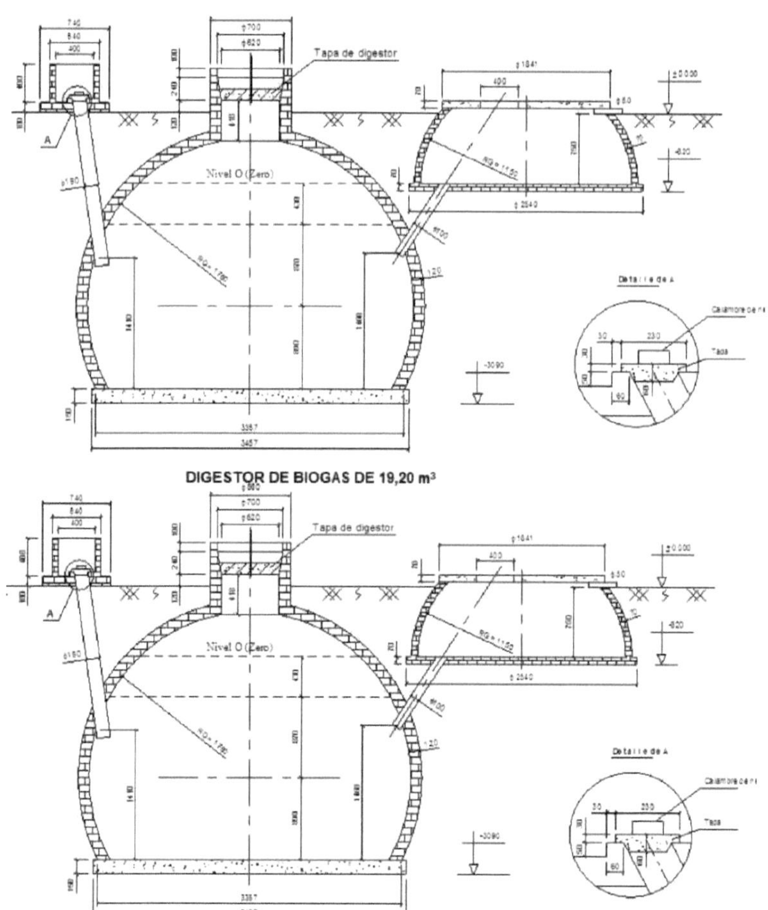

DIGESTOR DE BIOGAS DE 19,20 m³

DIGESTRO DE BIOGAS DE 19,58 m³

24 m

BIOGAS DIGESTER³

Tanque de regulación

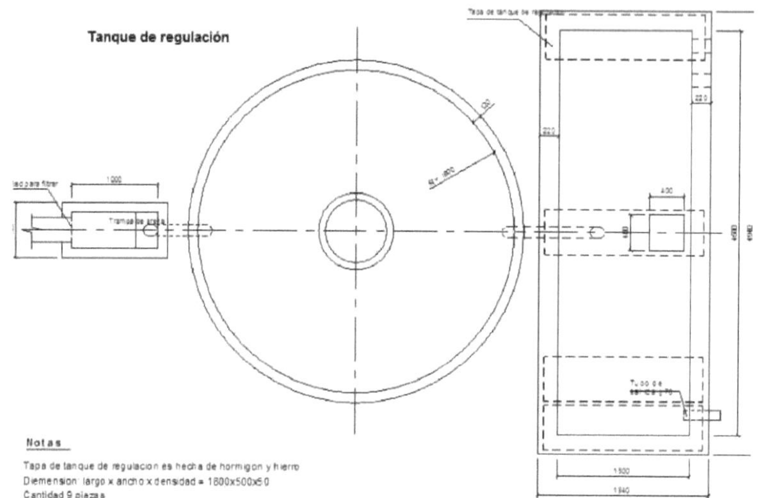

Notas

Tapa de tanque de regulacion es hecha de hormigon y hierro
Diemension: largo x ancho x densidad = 1800x500x50
Cantidad 9 piezas

Biodigester GBV 22 M^3 (Naum)

Estructura hierro en la tapa de tanque de regulacion
(9 piezas, 1 pieza con entrada para visitar)

Estructura hierro de tapa de recogedor de biogas
(Cantidad 1)

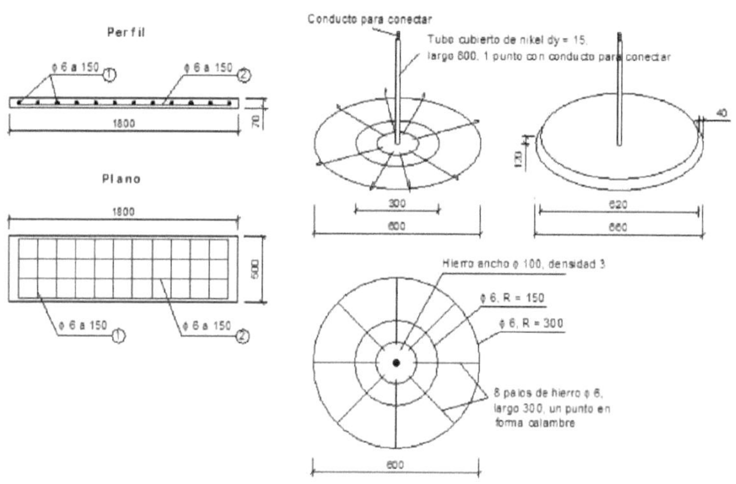

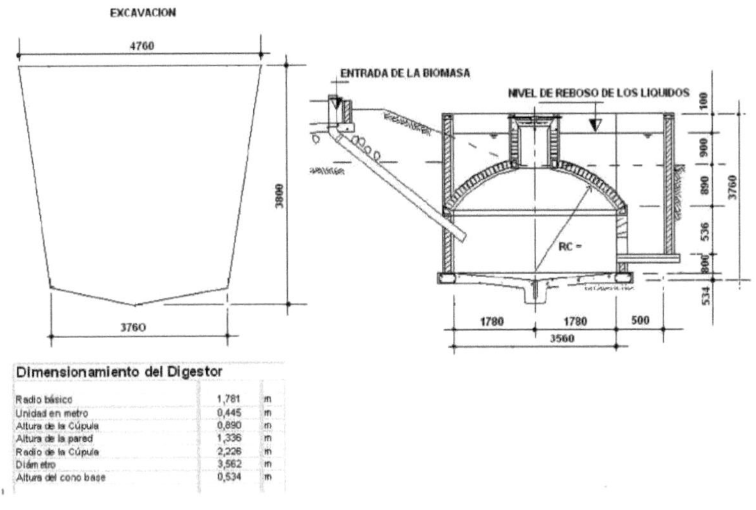

Dimensionamiento del Digestor

Radio básico	1,781	m
Unidad en metro	0,445	m
Altura de la Cúpula	0,890	m
Altura de la pared	1,336	m
Radio de la Cúpula	2,226	m
Diámetro	3,562	m
Altura del cono base	0,534	m

Biogas of 47 m3.

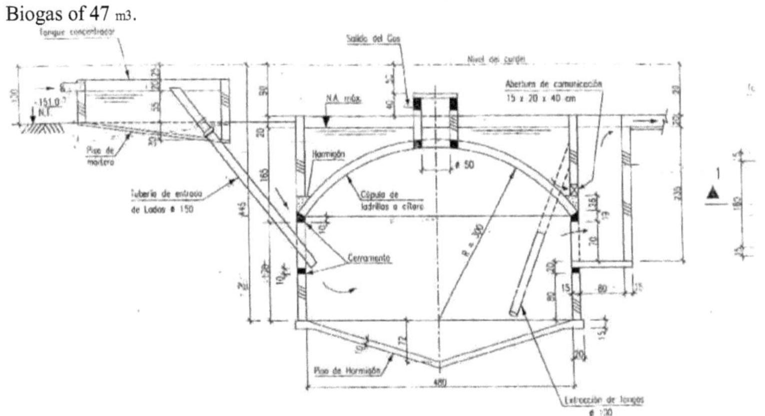

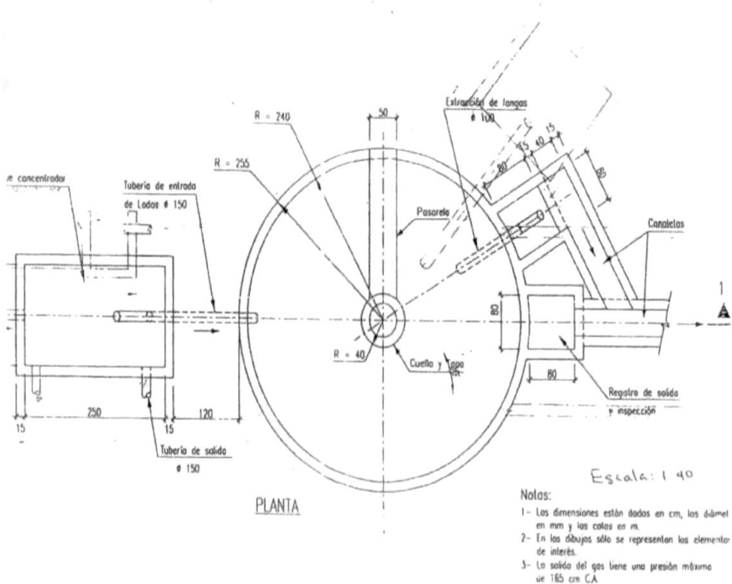
e concentrador
Tubería de entrada
de Lodos ⌀ 150
R = 240
R = 255
50
Extracción de langas
⌀ 100
15
15 40
80
80
Pasarele
Canaletas
R = 40
Cuello y Tapa
80
Registro de salida
y inspección
250
120
15
15
Tubería de salida
⌀ 150
PLANTA
Escala: 1 40
Notas:
1- Las dimensiones están dadas en cm, los diámet
en mm y las cotas en m.
2- En los dibujos sólo se representan los elemento
de interés.
3- La salida del gas tiene una presión máxima
de 165 cm C.A.

I **want** morebooks!

Buy your books fast and straightforward online - at one of world's fastest growing online book stores! Environmentally sound due to Print-on-Demand technologies.

Buy your books online at
www.morebooks.shop

Kaufen Sie Ihre Bücher schnell und unkompliziert online – auf einer der am schnellsten wachsenden Buchhandelsplattformen weltweit! Dank Print-On-Demand umwelt- und ressourcenschonend produziert.

Bücher schneller online kaufen
www.morebooks.shop

Printed by Books on Demand GmbH, Norderstedt / Germany